Progress in Numerical Simulation for Microelectronics Vol. 3

Edited by
K. Merten, Siemens AG, Munich
A. Gilg, Siemens AG, Munich

Alfred Kersch
William J. Morokoff

Transport Simulation
in Microelectronics

1995

Birkhäuser Verlag
Basel · Boston · Berlin

The Authors:

Dr. Alfred Kersch
Siemens AG
Otto-Hahn-Ring 6
D-81739 München, Germany

Dr. William J. Morokoff
Department of Mathematics
University of California
Los Angeles, CA 90024, U.S.A.

A CIP catalogue record for this book is available from the Library of Congress, Washington D.C., USA

Deutsche Bibliothek Cataloging-in-Publication Data
Kersch, Alfred:
Transport simulation in microelectronics / Alfred Kersch ; William J. Morokoff. -
Basel ; Berlin ; Boston : Birkhäuser, 1995
   (Progress in numerical simulation for microelectronics; Vol. 3)
   ISBN-13: 978-3-0348-9898-0   e-ISBN-13: 978-3-0348-9080-9
   DOI: 10.1007/978-3-0348-9080-9
NE: Morokoff, William J.:; GT

© 1995 Birkhäuser Verlag, P.O. Box 133, CH-4010 Basel, Switzerland
Camera-ready copy prepared by the authors
Printed on acid-free paper produced from chlorine-free pulp
Softcover reprint of the hardcover 1st edition 1995
ISBN-13: 978-3-0348-9898-0

987654321

Introduction

Computer simulation of semiconductor processing equipment and devices requires the use of a wide variety of numerical methods. Of these methods, the Monte Carlo approach is perhaps most fundamentally suited to modeling physical events occurring on microscopic scales which are intricately connected to the particle structure of nature. Here physical phenomena can be simulated by following simulation particles (such as electrons, molecules, photons, etc.) through a statistical sampling of scattering events. Monte Carlo is, however, generally looked on as a last resort due to the extremely slow convergence of these methods. It is of interest, then, to examine when in microelectronics it is necessary to use Monte Carlo methods, how such methods may be improved, and what are the alternatives. This book addresses three general areas of simulation which frequently arise in semiconductor modeling where Monte Carlo methods play a significant role. In the first chapter the basic mathematical theory of the Boltzmann equation for particle transport is presented. The following chapters are devoted to the modeling of the transport processes and the associated Monte Carlo methods. Specific examples of industrial applications illustrate the effectiveness and importance of these methods.

Two of these areas concern simulation of physical particles which may be assigned a time dependent position and velocity. This includes the molecules of a dilute gas used in such processing equipment as chemical vapor decomposition reactors and sputtering reactors. We also consider charged particles moving within a semiconductor lattice. In both cases various forms of the Boltzmann equation describe the time evolution of a probability distribution function for the position and velocity of a particle. It is this equation that the Monte Carlo particle simulations attempt to solve. Analysis of the Boltzmann equation indicates when other, faster numerical methods are suitable and suggests possible hybrid approaches which make limited use of Monte Carlo.

The third topic addressed here is radiative transfer which arises in Rapid Thermal Processing reactors, as well as many other areas. Here the simulated particles are photons which may be scattered or absorbed and thereby transfer energy from a source to a background or surface. Here the

Boltzmann equation may also be used as the mathematical model; however, now the speed of the particles is fixed at the speed of light, and only the velocity direction varies. We present here a less widely known treatment of the problem using Quasi-Monte Carlo methods whereby random sequences are replaced by more uniformly distributed deterministic sequences. This approach is particularly well suited to the linear, time-independent problems which often arise in radiative transport.

All of the key aspects of Boltzmann transport (non-linear scattering, linear scattering off a background and surface scattering) are addressed in the examples given. Numerous analogies may draw to other applications. For example, the simulation of small scale features of semiconductor device structures is closely related to radiation transport simulation. Ion implementation problems may be solved with methods similar to those used in the simulation of the sputtering reactor. And plasma simulation may be treated with a combination of methods used for gas dynamics and charge transport.

This book is the result of two different directions of research - applied mathematics and physical modeling - being applied to the area of numerical simulation of transport phenomena on the level of the Boltzmann equation. From the mathematical side, William Morokoff, working in a theoretically oriented, academic research environment, has attempted to unite these seemingly diverse areas through the similarities in their underlying mathematical models and to present a rigorous foundation for the numerical methods used. From the physical modeling side, Alfred Kersch, working in the context of an industrial research lab, has focused on presenting problems and examples of true industrial relevance and connecting these with the theoretically validated simulation methods. The authors have tried, through a close collaboration, to close the gap which often exists between the two approaches.

We greatfully acknowledge the opportunity given by the editors Knut Merten and Albert Gilg to include this book in their series "Progress of Numerical Simulation in Microelectronics (PNSM)". We would also like to acknowledge the cooperation of the people of the Siemens equipment simulation group, especially Chr.Werner and R.P. Brinkmann. Furthermore we thank Th. Vogelsang for the many things we learned from him and his PhD thesis, and Hermann Jacobs for his encouragement and support.

The collaboration between the authors had its origins in a joint program between Siemens and the Institute for Mathematics and Its Applications (IMA) at the University of Minnesota. We therefore give special thanks to the IMA and its director Avner Friedman for making this project possible. Thanks also to Russel Caflisch of UCLA for his support and advice, as well as to the National Science Foundation. Finally we wish to acknowledge the Universities of Minnesota, Arizona, California (Los Angeles) and L'Aquila (Italy) for the use of their resources in completing this project.

Contents

Chapter 1

The Boltzmann Equation

Systems consisting of a large number of particles are often most easily described by a probability density function $f(x, p, t)$, which gives the probability of finding a particle (molecule, photon, electron, etc.) at a position x and with momentum p. Macroscopic quantities such as mass density and temperature are found as averages over momentum space of the function f. It is then of interest to determine an equation for the time and space evolution of such distribution functions. When the evolution includes effects due to particle collisions, such an equation is called a Boltzmann equation. In this chapter we discuss the Boltzmann equation for rarefied gases, radiation and electrons, including a sketch of the derivation and the underlying assumptions.

1.1   The Liouville Equation

The starting point for the derivation of the Boltzmann equation is the Liouville equation, which describes the evolution of the probability density function for finding a system of particles at a given point in phase space. Therefore we give here a physical derivation of the Liouville equation starting from the equations of individual particle dymanics. Depending on the nature of the particle interactions and forces acting on the particles, one obtains different equations describing the evolution of a dilute gas or of electrons in a solid. We start by considering a system of N particles with positions and momenta $(x_i, p_i) \in D \times \mathbf{R}^3$ for $i = 1 \ldots N$, where D is a subset of $\mathbf{R}^3$ describing position space. Given the form of the force acting on each. particle, in principle it would be possible to determine the state of the system at any time after initial conditions are specified. This is exceptionally impractical, however, as the number of particles N is on the order of 10^{20}. Moreover, we are often only interested in macroscopic properties of the microscopic system such as temperature or heat transfer rates, which are averages over momentum space. As knowledge of the initial state is also likely to be inexact, we begin by considering the evolution of a probability density function on $6N$ dimensional phase space

$$P^N(\vec{x}, \vec{p}, t) = P^N(x_1, \cdots, x_N, p_1, \cdots, p_N, t) \tag{1.1}$$

such that $P^N(\vec{x}, \vec{p})\, d\vec{x}\, d\vec{p}$ is the probability of finding the system in a volume $d\vec{x}\, d\vec{p}$ around the point $(\vec{x}, \vec{p})$ in phase space. The initial data are chosen so that only states which correspond to specified initial macroscopic properties have non-zero probability.

The motion of the particles is described by the equations

$$\begin{aligned} \dot{\vec{x}} &= \vec{v} \\ \dot{\vec{p}} &= \vec{F} \end{aligned} \tag{1.2}$$

where the dot indicates time differentiation and $\vec{F}$ is the force (external and generated by the particles themselves). The particle velocity v_i is a function of the particle mass and momentum. For classical particles (including gas molecules) $v_i = p_i/m_i$. If a point in phase space is denoted by $\vec{z} = (\vec{x}, \vec{p})$, then Equations (1.2) may be written

$$\dot{\vec{z}} = \vec{Z} \tag{1.3}$$

where $\vec{Z} = (\vec{v}, \vec{F})$. If the forces are independent of momentum, as is usually the case (and even sometimes if they are not, as in the case of the Lorentz force on charged particles in a magnetic field), it follows that

$$\operatorname{div} \vec{Z} = \nabla_x \cdot \vec{v} + \nabla_p \cdot \vec{F} = 0 \ . \tag{1.4}$$

Liouville's theorem [20] then states that

$$\int_{\Omega(t)} d\vec{z} = \int_{\Omega(0)} d\vec{z_0} \ , \tag{1.5}$$

which is just conservation of volume $Omega$ in phase space under a divergence free mapping. From this it follows that the probability density function P^N satisfies

$$\frac{\partial P^N}{\partial t} + \vec{Z} \cdot \nabla P^N = 0 \tag{1.6}$$

which simply says that the probability that the system is in state $\vec{z}$ at time t is just the probability of being in state $\vec{z_0}$ at time 0 assuming that $\vec{z}$ satisfies Equations (1.3) and (1.4).

Equation (1.6) for P^N as a function of $6N$ variables is not any easier to solve than the original set of Equations (1.2). In general, however, we are not interested in the probability that all N particles are in a given state, but in the probability of finding one particle at a given point in 6 dimensional position-momentum space, independent of where the other particles are. Because all particles are considered indistinguishable (appropriate modifications must be made for mixtures), the one particle distribution function is the same for all N particles and may be expressed in terms of P^N through the formula

$$P_N^1(x, p, t) = \int P^N \prod_{i=2}^{N} dx_i dp_i \qquad (1.7)$$

The goal is then to derive an equation for $P_N^1(x, p, t)$. It should be noted that without further assumptions on P^N, a useful equation could not be obtained. This is because there are many different functions $P^N(t = 0)$ which integrate to give the same $P_N^1(t = 0)$. As these different initial functions P^N evolve according to (1.6), the corresponding P^1's will become quite different. The assumption of "molecular chaos", described below, allows this problem to be overcome.

Another question which should be addressed is how an equation for $P_N^1(x, p, t)$ could describe a collection of particles approaching equilibrium when it is derived from the Liouville equation (1.6), which is time reversible. Mathematically, this difficulty is avoided by considering an equation for the one particle distribution function while allowing N to go to infinity. Conservation of volume in an infinite dimensional space is then no longer at odds with an equilibrium state. In order to make physical sense it is necessary that the particle size go to zero so that the total volume occupied by the N particles in the limit remains finite. If a particle has diameter d, then the volume occupied by the particles is of order Nd^3. An ideal gas is one such that $Nd^3 \to 0$. In this case the particles occupy a negligible fraction of the total volume. If it is also true that $Nd^2 \to 0$, then collisions between particles are said to be negligible, and the distribution function is influenced only by boundary interactions, external forces, and weak self-consistent forces which describe the effect of all the other particles collectively on a particle in question. This assumption is frequently made when dealing with electrons. When the limit of Nd^2 is non-zero, then binary collisions provide the mechanism for the approach to equilibrium. This is the case of rarefied gases in the so called transition regime between free molecular flow (no collisions) and a continuum, where the gas is so dense that particle nature may be disregarded.

1.2 Rarefied Gases

The Boltzmann equation will now be considered for a collection of molecules in a gas. The following is based on the rigorous and detailed development presented in [20]. For simplicity the gas is assumed to consist of a single species of monatomic molecules which interact pair-wise according to a potential $\Phi(r)$, where r is the distance between the molecules. To model a dilute gas, where only binary collisions are important, the potential is set to zero for distances r greater than a cut-off value d, which is regarded as the molecular diameter. We will assume that no external forces are present, so that the only forces exerted on a molecule are due to short range binary collisions. As usual, the classical particle assumption is made, so that the equations are presented in terms of the particle velocity v (as opposed to the momentum p).

The Boltzmann equation is usually written in terms of mass density, defined by

$$f^1(x, v, t) = \lim_{N \to \infty} N m P_N^1(x, v, t) . \tag{1.8}$$

Here m is the mass of a molecule, which is assumed to go to zero as N goes to infinity, so that the total mass Nm remains finite. If the Liouville equation (1.6) is integrated with respect to the variables x_i, v_i for $i = 2 \ldots N$, in the limit $N \to \infty$ we obtain

$$\frac{\partial f^1}{\partial t} + v \cdot \nabla_x f^1 = \int \int \int [f^2(v', w') - f^2(v, w)] \, B(V_r, \chi)/m \, d\chi \, d\psi \, dw .$$
$$\tag{1.9}$$

Here the integration on the right hand side is over $0 \leq \chi \leq \pi$ and $0 \leq \psi \leq 2\pi$ and over all of velocity space $w \in \mathbf{R}^3$. The quantity V_r is the relative velocity of a colliding pair, defined by $V_r = |v - w|$, and v' and w' are post collision velocities related to v and w by

$$v' = v_m + \frac{1}{2} V_r \, \vec{n}' \tag{1.10}$$

$$w' = w_m - \frac{1}{2} V_r \, \vec{n}' .$$

Here $v_m = (v + w)/2$ is the center of mass velocity. The unit vector $\vec{n}'$ is the post-collision direction of the relative velocity defined in terms of the pre-collision direction of the relative velocity $\vec{n}$ by

$$\vec{n}' = R(\vec{n})\vec{q} . \tag{1.11}$$

Here $\vec{q}$ is given by

$$\vec{q} = (\cos \chi, \, \sin \chi \, \cos \psi, \, \sin \chi \, \sin \psi) . \tag{1.12}$$

The unit vector $\vec{q}$ is the direction of the post-collision relative velocity in the reference frame in which the pre-collision relative velocity has coordinates $(1, 0, 0)$. Thus a value of $\chi = 0$ corresponds to a minimally glancing collision (i.e., no change in the direction of the relative velocity), while a value of $\chi =$

π corresponds to a head-on collision resulting in a reversal in direction (i.e., $\vec{n}' = -\vec{n}$). It is of course necessary to map the vector $\vec{q}$ into the center of mass reference frame; thus the presence of the orthonormal rotation matrix $R(\vec{n})$ in Equation (1.11). This matrix is a non-linear function of the components of $\vec{n}$ and has the property that it maps the vector $(1, 0, 0)$ to $\vec{n}$. The mapping of the vector $(0, 1, 0)$ or $(0, 0, 1)$ is however arbitrary because the scattering depends only on the angle χ, but not on the orientation of the scattering plane given by ψ. Thus $R(\vec{n})$ is not unique; a possible choice for R is given explicitly in Section 3.1.

$B(V_r, \chi)$ describes the nature of the molecular collisions and depends on the potential $\Phi(r)$. This relationship is discussed below. Physically, $B(V_r, \chi)$ is the volume, per unit time, of a region of physical space containing particles which will collide with a given particle with relative velocity V_r and scatter at an angle χ. The quantity $B(V_r, \chi)f(w)$ may be regarded as the unnormalized probability of finding such a collision partner. Here it is assumed that the molecular scattering does not depend on internal degrees of freedom such as rotation and vibration; this is why $B(V_r, \chi)$ is independent of the orientation of the scattering plane.

The only term of Equation (1.9) which has not been identified is $f^2(v, w)$. This is the two particle mass density defined by

$$f^2(x_1, v_1, x_2, v_2, t) = \lim_{N \to \infty} (Nm)^2 \int P^N \prod_{i=3}^{N} dx_i dv_i . \qquad (1.13)$$

The quantity $f^2(v, w)/m^2$ gives the number density of particle pairs (at position x and time t) such that one particle has velocity v and the other has velocity w. The integral term of Equation (1.9) then gives the net change of $f^1(v)$ due to collisions, with $f^2(v', w')$ representing the particle pairs which collide to produce a particle with velocity v (thus increasing $f^1(v)$), while $f^2(v, w)$ describes particles which collide and scatter out of velocity v, thereby decreasing $f^1(v)$ (hence the minus sign).

The fact that f^1 is determined from f^2 embodies the difficulty mentioned above of finding a single equation for f^1 without further assumptions on the form of P^N. Equation (1.9) is in fact only the first of a hierarchy of equations for f^s, $s = 1, 2, \ldots$, involving an integral of f^{s+1}. It is necessary then to find an assumption on f^s which is consistent within the hierarchy such that a single equation for f^1 may be obtained. This assumption is known as molecular chaos, which means that the distribution functions satisfy the factorization property

$$f^s = \prod_{i=1}^{s} f^1(x_i, v_i, t) . \qquad (1.14)$$

In particular the two particle density function may be expressed as

$$f^2(x_1, v_1, x_2, v_2, t) = f^1(x_1, v_1, t)\, f^1(x_2, v_2, t)\ . \tag{1.15}$$

This can be interpreted to mean that the probability of finding a pair of molecules with given positions and velocities is just the product of the probability of finding each molecule with the specified position and velocity. This seems quite plausible for a pair of randomly chosen particles, and also for pairs about to collide. It should be noted that Equation (1.9) was derived such that f^2 appears only for particles before a collision. There remains the problem, however, that for particles which have just collided, the assumption cannot be true. It may be true in the limit $Nd^3 \to 0$, though, for as the molecular size and interaction time goes to zero, the size of the set for which molecular chaos does not hold may become negligible. In addition it is possible to show that in this limit if the initial data for the hierarchy of equations satisfy this assumption, then the solution at any finite time will also satisfy the factorization property. Thus the assumption of molecular chaos, which allows for a single closed equation for f^1, is at least consistent with the original equations.

The result of factorizing f^2 is the Boltzmann equation. As we are now only interested in the one particle density function, we will drop the superscript and write f in place of f^1 so that the equation becomes

$$\frac{\partial f}{\partial t} + v \cdot \nabla_x f = \frac{1}{m} \int \int \int \left[f(v')f(w') - f(v)f(w) \right] B(V_r, \chi)\, d\chi\, d\psi\, dw\ .$$
$$\tag{1.16}$$

This is a non-linear integro-differential equation for a function of seven variables which includes a five dimensional integral over an unbounded domain. As discussed below, in certain limits this equation may be reduced to the Euler or Navier-Stokes equations for continuum flow. However, in general, the description of the evolution of a rarefied gas requires the solution of the full Boltzmann equation. Difficulties with discretizing velocity space and evaluating a five dimensional integral at each point in physical space strongly favor the use of a Monte Carlo method to obtain a numerical solution.

Consider for the moment the case where d and m are non-zero (i.e., the limit $N \to \infty$ has not yet been taken). The molecular interaction term $B(V_r, \chi)$ may be derived from considering the mechanics of a binary collision. The assumption of a binary short range interaction due to a spherical potential means that the only force acting on a particle comes from its collision partner, and that this force is directed along the line which connects their centers. It follows from conservation of linear momentum that the vector connecting the centers and the relative velocity vector both lie in the same plane, which has constant orientation in time. Moreover, this collision plane moves with constant velocity equal to the center of mass velocity. Thus the entire collision dynamics may be represented in this plane, whose

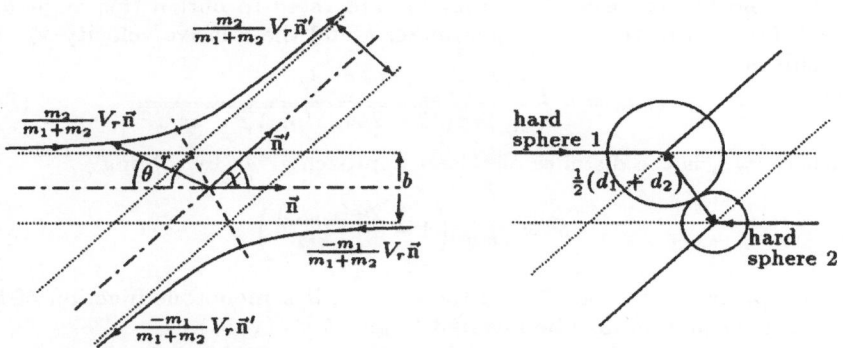

Figure 1.1: The geometry of the scattering interaction in the center of mass reference frame and the special case of the interaction of spheres with diameter d_1 and d_2, respectively.

orientation is specified by the angle ψ above. The dynamics are determined by the conservation of energy and angular momentum. It turns out that the resulting equations are identical to the equations for a single particle with mass given by the reduced mass of the system, which is defined as

$$m_r = \frac{m_1 m_2}{m_1 + m_2} \,, \tag{1.17}$$

moving through the potential field $\Phi(r)$ with velocity given by the relative velocity $V_r = v - w$. If particles are identical (as is our case), then the reduced mass is simply $m/2$. This interaction is sketched in Figure 1.1. In polar coordinates the distance from the center of the potential to the particle is r and the angle is θ. The distance of closest approach of the particles assuming no interaction (that is, the potential is zero everywhere) is known as the impact parameter b. The conservation of energy and angular momentum in this reference frame are then

$$\frac{1}{2} m_r \left(\dot{r}^2 + r^2 \dot{\theta}^2 \right) + \Phi(r) = \frac{1}{2} m_r V_r^2 \tag{1.18}$$

$$r^2 \dot{\theta} = b V_r$$

These equations are valid for $r \leq d$, where $\Phi(r) = 0$ for $r > d$. Again the dot indicates a time derivative and V_r is the magnitude of the precollision (as well as post-collision) relative velocity.

By symmetry, the point of closest approach during the collision determines a unique angle θ_{min}. This is related to the scattering angle χ in Equation (1.10) by

$$\chi = \pi - 2\,\theta_{min} \,. \tag{1.19}$$

The conservation laws above may be integrated to obtain this angle θ_{min} as a function of the impact parameter b and the relative velocity V_r. The result is

$$\theta_{min} = b \int_{r_{min}}^{\infty} \frac{dr}{r^2 \sqrt{1 - 2\Phi(r)/m_r V_r^2 - b^2/r^2}} \qquad (1.20)$$

where r_{min} is the distance of closest approach given by solving

$$b^2 = r_{min}^2 \left(1 - \frac{2\Phi(r_{min})}{m_r V_r^2} \right) . \qquad (1.21)$$

For repulsive potentials θ, and therefore χ, is a monotone function of b so that this equation may be inverted to give $b = b(\chi, V_r)$.

It remains to be seen what is the function $B(V_r, \chi)$. In the derivation of the Boltzmann equation this quantity represents the volume (per unit time) in physical space which contains particles with the given precollision relative velocity which will collide and scatter with the given angle. This volume can be expressed as

$$B(V_r, \chi)d\chi d\psi = V_r\, b\, db\, d\psi \qquad (1.22)$$

where again b is the impact parameter and V_r is the relative speed. From the analysis of the collision dynamics we have $b = b(\chi, V_r)$ so that we obtain

$$B(V_r, \chi)d\chi d\psi = V_r\, b \left| \frac{\partial b}{\partial \chi} \right| d\chi d\psi \qquad (1.23)$$

Specifying the molecular potential $\Phi(r)$ then determines $B(V_r, \chi)$. For example, for the case of hard spheres, where $\Phi(r) = \infty$ for $r < d$, it follows that

$$B(V_r, \chi) = \frac{1}{4}V_r d^2 \sin \chi . \qquad (1.24)$$

The quantity

$$d\sigma(b, \chi) = b \left| \frac{\partial b}{\partial \chi} \right| d\chi d\psi = \frac{b}{\sin \chi} \left| \frac{\partial b}{\partial \chi} \right| d(\cos \chi)d\psi \qquad (1.25)$$

is known as the differential scattering cross section. Integrals of this quantity have various physical interpretations. This is discussed further in Section 2.4.1.

Passing to the limit $N \to \infty$, $d \to 0$ and $m \to 0$ does not present any problem because the quantity $B(V_r, \chi)/m$ effectively behaves like d^2/m, which is finite. The limiting value may be represented by $B(V_r, \chi)/m$ where now m is the actual mass of a physical atom and $B(V_r, \chi)$ is a function of the actual molecular diameter (i.e., the cut off parameter for the interaction potential).

1.2.1 Equilibrium

The Boltzmann equation is frequently written

$$\frac{\partial f}{\partial t} + v \cdot \nabla_x f = Q(f, f) \tag{1.26}$$

where $Q(f, f)$ is the collision integral given in Equation (1.16). We consider for the moment the spatially homogeneous case and ask about solutions that do not change with time. The result is that any velocity distribution function satisfying

$$Q(f, f) = 0 \tag{1.27}$$

must necessarily be a Maxwellian distribution of the form

$$f(v) = \frac{\rho}{(2\pi RT)^{3/2}} \exp\left(\frac{-(v - v_0)^2}{2RT}\right) . \tag{1.28}$$

Here ρ, v_0 and T are constants identified with mass density, stream velocity and temperature. $R = k_B/m$ is the gas constant, where k_B is the Boltzmann constant and m is the molecular mass. What's more, it can be shown that any solution of the spatially homogeneous time dependent equation must tend toward this equilibrium solution.

In order to obtain this result we begin by considering the quantity

$$\int \phi(v) Q(f, f) \, dv \tag{1.29}$$

for any function $\phi(v)$ on velocity space such that this integral exists for all solutions of Equation (1.26). Repeated use of the reversibility of individual collisions, Equation (1.10), and the fact that $B(V_r, \chi)$ depends only on the relative speed, which is the same before and after the collision, allows the quantity of interest to be expressed in four different ways through change of variables among the velocities v, w, v' and w'. When these equations are added the result is

$$\int \phi(v) Q(f, f) dv = \frac{1}{4} \int \int \int \int [\phi(v) + \phi(w) - \phi(v') - \phi(w')] \tag{1.30}$$
$$\times [f(v')f(w') - f(v)f(w)] B(V_r, \chi) \, d\chi d\psi dw dv .$$

Because $Q(f, f)$ describes changes which occur as the results of collisions, any function $\phi(v)$ which satisfies

$$\int \phi(v) Q(f, f) \, dv = 0 \tag{1.31}$$

for all solutions f of the Boltzmann equation is called a collision invariant. This will be true if and only if

$$\phi(v) + \phi(w) = \phi(v') + \phi(w') \tag{1.32}$$

for all velocity pairs related through Equation (1.10) for any $\vec{n}'$. It is possible to show that this algebraic relationship is satisfied only by functions of the form

$$\phi(v) = a + b \cdot v + cv^2 \,, \tag{1.33}$$

which is to say that the only collision invariants are $1, v$ and v^2 corresponding to mass, momentum and energy.

If $\phi(v) = \log f(v)$ is substituted into Equation (1.30), then this equation becomes

$$\int \phi(v)Q(f,f)dv = \frac{1}{4} \int\int\int\int f(v')f(w')[1-\alpha]\log\alpha B(V_r,\chi)d\chi d\psi dw dv, \tag{1.34}$$

where $\alpha = f(v)f(w)/f(v')f(w') \geq 0$. It is easily seen that

$$[1-\alpha]\log\alpha \leq 0 \tag{1.35}$$

for any non-negative α and equality holds only when $\alpha = 1$. Therefore it follows that

$$\int \log f(v)Q(f,f)\ dv \leq 0 \,, \tag{1.36}$$

with equality if and only if $f(v)f(w) = f(v')f(w')$. This statement along with the definition of Q allows us to conclude that $Q(f,f) = 0$ if and only if $f(v)f(w) = f(v')f(w')$. But by taking the logarithm of both sides we find that $\phi(v) = \log f(v)$ must satisfy Equation (1.32). Therefore $Q = 0$ if and only if $f(v) = \exp(a + b \cdot v + cv^2)$ for some constants a, b and c. Equation (1.28) follows from some minor algebraic manipulations and redefinition of constants.

The tendency to equilibrium in the spatially homogeneous case can be seen by investigating the function

$$H(t) = \int f(v,t)\log f(v,t)\ dv \,, \tag{1.37}$$

which is interpreted as a kind of negative entropy. The time derivative of $H(t)$ satisfies

$$\begin{aligned} \frac{dH}{dt} &= \frac{d}{dt}\int f(v,t)\log f(v,t)\ dv \\ &= \int (1+\log f)\frac{\partial f}{\partial t}\ dv \\ &= \int (1+\log f)Q(f,f)\ dv \\ &= \int \log f Q(f,f)\ dv \leq 0 \,. \end{aligned} \tag{1.38}$$

Here the fact that 1 is a collision invariant was used. This shows that H is a decreasing function which tends towards a Maxwellian equilibrium. This is known as the H-theorem. It may be extended to the non-homogeneous case, but then boundary data must be included. For general boundary conditions, one does not expect that the steady state will be in thermodynamic equilibrium.

1.2.2 Knudsen Number and Fluid Equations

A great deal can be learned about the Boltzmann equation by considering certain limits of the non-dimensional equation. If the equation is scaled using parameters which include a typical macroscopic length $\mathcal{L}$ and a typical number density n, and the term $B(V_r, \chi)$ (which has units of area $\times$ velocity) is scaled by d^2 times a typical molecular velocity, then the resulting non-dimensional equation appears as

$$\frac{\partial f}{\partial t} + v \cdot \nabla_x f = \frac{1}{\epsilon} Q(f, f) . \tag{1.39}$$

Here ϵ is defined by

$$\epsilon = \sqrt{2}\pi \left(\frac{\lambda}{\mathcal{L}} \right) \tag{1.40}$$

where λ is the hard sphere mean free path defined as

$$\lambda = \frac{1}{\sqrt{2}\pi d^2 n} . \tag{1.41}$$

The dimensionless quantity $Kn = \lambda/\mathcal{L}$ is known as the Knudsen number. It compares the length scale over which microscopic events occur to the length scale for macroscopic events. From the equation it can be seen that in the limit $\epsilon \to 0$, the collision term dominates and the distribution function quickly approaches a local equilibrium. A particle undergoes a great number of collisions before moving to a position which is macroscopically different from its origin. This situation describes a continuum gas, and thus $\epsilon \to 0$ is said to be the continuum limit. At the other extreme, $\epsilon \to \infty$, interparticle collisions become insignificant and the flow is determined solely by the boundary interactions. This is the free molecular flow limit.

It should be mentioned that for many real problems there is not a unique Knudsen number. The characteristic length scale may vary from region to region, and the mean free path, which depends on density, is by no means constant. Frequently there are both continuum and rarefied regions in the same problem. It is perhaps better then to speak of a local Knudsen number. Equation (1.39) remains mathematically sound for constant ϵ; however the interpretation of ϵ as a physical quantity must be carefully considered.

When the Boltzmann equation is integrated against the collision invariants, the result is a system of conservation laws for mass, momentum and energy similar to those for continuum fluid mechanics.

$$\frac{\partial \rho}{\partial t} + \nabla \cdot (\rho v_0) = 0 \tag{1.42}$$

$$\frac{\partial v_0}{\partial t} + v_0 \cdot \nabla v_0 = -\frac{1}{\rho} \nabla \cdot P$$

$$\frac{3}{2} \rho \frac{k_B}{m} \left(\frac{\partial T}{\partial t} + v_0 \cdot \nabla T \right) = -\nabla \cdot q - P : \nabla v_0$$

Here the quantities ρ, v_0 and T, identified as mass density, stream velocity and temperature, are defined by

$$\rho(x, t) = \int f \, dv \tag{1.43}$$

$$v_0(x, t) = \frac{1}{\rho} \int v f \, dv$$

$$T(x, t) = \frac{m}{3 k_B \rho} \int (v - v_0)^2 f \, dv \ .$$

Again, m is the mass of a molecule and k_B is the Boltzmann constant. The quantity q is the heat flux vector defined as

$$q(x, t) = \frac{1}{2} \int (v - v_0)^2 (v - v_0) f \, dv \tag{1.44}$$

and the two dimensional symmetric stress tensor P is given, component-wise, as

$$P_{ij} = \int (v - v_0)_i (v - v_0)_j f \, dv \tag{1.45}$$

From these definitions one can deduce one further relationship, the equation of state

$$p = \frac{1}{3} \sum_{i=1}^{3} P_{ii} = \rho \left(\frac{k_B}{m} \right) T \ . \tag{1.46}$$

In the case of equilibrium, p is the hydrostatic pressure and $P = pI$, where I is the identity matrix.

Equations (1.42) together with the equation of state form a system of 6 equations in 14 variables, which in general cannot be solved. However, in the continuum limit $\epsilon \to 0$ the small parameter ϵ may be used as the basis for an asymptotic expansion of f. Closure of this system may then be obtained by truncating the expansion. One possible expansion, known as the Hilbert expansion, is of the form

$$f(x, v, t) = f^0 + \epsilon f^1 + \epsilon^2 f^2 \cdots \tag{1.47}$$

where the f^i are taken to be independent of ϵ. Substitution into Equation (1.39) shows that the leading order term must be a Maxwellian of the form in Equation (1.28). If the series is truncated at this term, so that $f = f^0$, the stress tensor reduces to the equilibrium value $P = pI$ and the heat flux vector q is zero. The system (1.42) then becomes the Euler equations for compressible, inviscid flow.

One difficulty with this approach is that no matter where the series is truncated, the corresponding fluid variables determined by Equations (1.42) will never be the exact ρ, v_0 and T determined by the true solution f. This may be remedied by requiring that all the f^i with $i \geq 1$ be orthogonal to the collision invariants, i.e. $\int \psi f^i dv = 0$ for $\psi = 1$, v or v^2. Then all of the fluid dynamic information is contained in f^0. Again from Equation (1.39) f^0 must be a Maxwellian, but now ρ, v_0 and T depend on ϵ. This is the Chapman-Engskog expansion. When the first order term of the expansion is included, the analysis becomes rather lengthy, but the result is that the system (1.42) is the Navier-Stokes equations. In addition, under some mild assumptions, for inverse power law potentials ($\Phi(r) = r^{-\eta}$ with $\eta \geq 4$) the viscosity and heat conduction coefficients can be shown to have temperature dependence of the form

$$\mu(T) = C\, T^\omega \qquad (1.48)$$

where ω depends on η and is between 0.5 and 1. Moreover, boundary conditions may be derived for the Navier-Stokes equation which represent the well known slip phenomena. For example, for a single species, isothermal gas flowing by a stationary wall, the proper boundary condition for the velocity at the wall ($z = 0$) is

$$v|_{z=0} = \zeta^v \frac{\partial v_t}{\partial n} \qquad (1.49)$$

Here $v_t = v - (v \cdot \vec{n})\vec{n}$ is the tangential velocity of the gas at the wall and $\partial/\partial n = \vec{n} \cdot \nabla$ is the derivative in the direction of the wall normal vector $\vec{n}$. The velocity slip coefficient ζ^v is proportional to mean free path so that as the gas becomes more dense, these effects go to zero. The coefficient also depends on the nature of the particle-wall interaction. Similar boundary conditions exist for temperature and for the phenomena of thermal creep and diffusive creep, which occurs for gas mixtures. In general, Equation (1.49) will also contain terms which are linear in temperature gradients and partial pressure gradients.

Further equations may obtained by using higher order terms in the Chapman-Engskog expansion. However, the resulting equations contain higher order derivatives in velocity which have limited physical interpretation. Also it is unclear what additional boundary conditions should be imposed. Thus only the first order expansion has been put to practical use.

1.2.3 Multiple Species

The above analysis may be extended to a gas mixture consisting of s different species by considering the evolution of a separate density function $f_k(x, v, t)$ for each species. These are normalized so that

$$\int f_k(x, v, t)\, dv = \rho_k(x, t) \tag{1.50}$$

is the mass density of the kth species. The restriction of pair-wise interactions allows the change of f_k due to collisions to be expressed as the sum over all species j of the collision term describing the interaction of species k with j. Thus we obtain equations for the s species, $k = 1, ..., s$, of the form

$$\frac{\partial f_k}{\partial t} + v \cdot \nabla_x f_k = \tag{1.51}$$

$$\sum_{j=1}^{s} \frac{1}{m_j} \int\int\int \left[f_k(v') f_j(w') - f_k(v) f_j(w) \right] B_{k,j}(V_r, \chi)\, d\chi d\psi dw.$$

Here the pre- and post-collision velocities v, w, v' and w' are no longer related through Equation (1.10) because the colliding particles are not necessarily of equal mass. The correct equations are

$$v' = v_m - \frac{m_r}{m_k} V_r\, \vec{n'} \tag{1.52}$$

$$w' = v_m + \frac{m_r}{m_j} V_r\, \vec{n'}$$

where m_r is the reduced mass of the colliding pair defined by Equation (1.17).

For most applications involving rarefied gas flows in microelectronics, the representation of the gas given above, as a mixture of non-reacting monatomic spherical molecules, is sufficient. While surface chemistry is of prime importance, the species and conditions present in a typical reactor indicate that gas-phase chemistry is of minor importance. Moreover, operating conditions generally involve temperatures low enough so that internal energy (rotational and vibrational) may be neglected. Therefore we will simply mention, without elaboration, that it is possible to modify Equation (1.51) to include these factors. The resulting equation for the s species, $k = 1, .., s$, has the form

$$\frac{\partial f_k}{\partial t} + v \cdot \nabla_x f_k = \tag{1.53}$$

$$\sum_{i,j,l=1}^{s} \frac{1}{m_l} \int\int\int \left[f_i(v') f_j(w') - f_k(v) f_l(w) \right] W_{k,l}^{i,j}(v, w, v', w') dw dv' dw'.$$

The velocities are now taken to be independent. The quantity $W_{k,l}^{i,j}$ is the transition probability that a molecule of species k with velocity v will react with a molecule of species l with velocity w to produce species i with velocity v' and species j with velocity w'. The specifics of the reactions and energy partitioning must be built into W, which will generally contain a delta function restricting the interactions to those which conserve momentum and energy.

1.2.4   Boundary Conditions

In order to complete the description of the time evolution of a rarefied gas, it is necessary to give boundary and initial conditions for the density function $f(x, v, t)$. In velocity space, the domain is usually taken to be all of $\mathbf{R}^3$. The only boundary condition for velocity v imposed on f is then that $f(v) \to 0$ sufficiently fast as $v \to \infty$ so that $v^2 f$ is integrable. For most problems of interest in microelectronics, it is the steady-state Boltzmann equation which is of interest, so the initial condition in time is not important. For those situations where a transient solution is desired, the initial conditions are taken to be the steady solution for a fixed set of spatial boundary conditions, and the time dependence is then introduced by temporally varying these boundaries.

The key question of interest is then how to describe the spatial boundary conditions. In the follozing we will assume that the boundaries are stationary, so that the wall velocity is zero. The extension to the case of moving boundaries is striaghtforward. The simplest model for gas-surface interaction is specular reflection. Here the surface is assumed to be perfectly smooth and flat at the point of contact x, so that the particle reflects instantaneously back into the gas from the same point with post-reflection velocity given by

$$v = w - 2(w \cdot n(x)) \, n(x) \, , \qquad (1.54)$$

where $n(x)$ is the inward normal vector to the surface at x. The boundary condition for $f(x, v, t)$ at a specularly reflecting surface is then just

$$f(x, v, t) = f(x, w, t) \, , \; (w \cdot n(x) < 0) \qquad (1.55)$$

with v given by Equation (1.54). In specular reflection the wall temperature does not influence the scattering.

The opposite extreme is an interaction whereby the particles striking the surface become 'captured' by the surface and lose all information about the distribution from which they came. They are subsequently re-emitted into the gas with a Maxwellian equilibrium distribution based on the wall temperature T_w. Assuming conservation of mass for boundary interactions, this may be expressed as

$$f(x, v, t) = \frac{1}{2\pi (RT_w)^2} \exp\left(\frac{-v^2}{2RT_w}\right) \int_{w \cdot n(x) < 0} |w \cdot n(x)| f(x, w, t) dw$$

$$(v \cdot n(x) > 0), \qquad (1.56)$$

where the normalization is chosen so that the mass per unit area per unit time leaving the surface is equal to that which arrives, i.e.

$$\int_{v \cdot n(x) > 0} |v \cdot n(x)| f(x, v, t) \, dv = \int_{v \cdot n(x) < 0} |v \cdot n(x)| f(x, v, t) \, dv . \quad (1.57)$$

Surfaces which satisfy this kind of law are called diffusive.

Specularly reflecting and diffusive surfaces can be considered special cases of a general boundary condition of the form

$$|v \cdot n(x)| f(x, v, t) = \int_{w \cdot n(x) < 0} R(w \to v, x, t) |w \cdot n(x)| f(x, w, t) dw$$

$$(v \cdot n(x) > 0), \qquad (1.58)$$

where $R(w \to v)$ is the transition probability that a molecule striking the surface with velocity w $(w \cdot n(x) < 0)$ will leave the surface with velocity v $(v \cdot n(x) > 0)$. R, as it represents a probability, is assumed to be non-negative,

$$R(w \to v, x, t) \geq 0 , \qquad (1.59)$$

to conserve mass,

$$\int_{v \cdot n(x) > 0} R(w \to v, x, t) \, dv = 1 , \qquad (1.60)$$

and to satisfy the principal of detailed balance:

$$|v \cdot n| f_0(v) R(-v \to -w) = |w \cdot n| f_0(w) R(w \to v) \qquad (1.61)$$

$$(v \cdot n > 0 , \ w \cdot n < 0) .$$

Here f_0 is a Maxwellian with zero mean velocity. Specular reflection corresponds to R being a delta function, and diffuse reflection corresponds to R being an appropriately scaled Maxwellian in v, independent of w.

If the proper physics can be identified, this may be built in to R to give realistic boundary conditions. The restriction of conservation of mass may be relaxed to allow for particle absorption. In this way surface coverage and surface reactions may be handled; in other applications, it is possible to model radiation or neutron absorption in this way. In the case of somewhat higher densities or surface absorption, it may be necessary to make R a function of f to account for the fact that particles are not scattering individually, independent of one another. Chemical reactions will also in general require that R_k for species k be a function of f_j for all the other species j. The main disadvantage of this approach, however, is that detailed knowledge of the physical and chemical behavior of particle-surface interactions is rarely available. Frequently little attempt is made to model the true physics; instead, a simple model with adjustable parameters is used, and the parameters are based on experiments for specific surfaces and gases.

The most common approach of this type is that of Maxwell's accommodation coefficient. In this case, it is postulated that a certain percentage α of the molecules striking the surface are captured and then re-emitted according to diffuse reflection. The remaining molecules scatter according to the laws of specular reflection. The quantity α is called the accommodation coefficient. The transition probability density function associated with this scattering law is then

$$
\begin{aligned}
R(w \to v, x, t) \;=\;& \alpha |v \cdot n| f_0(v) + (1 - \alpha)\delta(v - w + 2(w \cdot n(x))n(x)) \\
& (v \cdot n(x) > 0 \;,\; w \cdot n(x) < 0) \,.
\end{aligned}
\tag{1.62}
$$

Here f_0 is the zero velocity Maxwellian at the wall temperature, normalized as in Equation (1.56). The choice of the accommodation coefficient is then based on experimental data. It is also possible to develop more complicated mathematical models which involve multiple accommodation coefficients, which may include, for example, coefficients for tangential momentum and normal momentum. However, the detail which is included is often limited by the availability of data.

Open boundaries which allow inflow and outflow must also be considered. In this case the density function $f^+(x, v, t)$ is specified for each x on such a boundary and for all v such that $v \cdot n(x) > 0$. Molecules which cross the boundary from the interior disappear and do not influence the imposed f^+. Thus there is no conservation across open boundaries. For lack of any other information f^+ is generally taken to be a flux weighted Maxwellian of the form

$$
f^+(x, v, t) = C\, \rho\, v \cdot n(x)\, \exp\left(\frac{-(v - v_0)^2}{2RT}\right)
\tag{1.63}
$$

defined for $v \cdot n(x) > 0$. Here T and ρ are the external temperature and density, while v_0 is the stream velocity. C is a normalization factor defined by

$$
\frac{1}{C} = \int_{v \cdot n(x) > 0} \exp\left(\frac{-(v - u_0)^2}{2RT}\right) \,.
\tag{1.64}
$$

For typical microelectronic applications, v_0 will be fairly small and certainly subsonic. This leads to some difficulties, because the molecules that leave the domain should influence the exterior gas coming in. However, by specifying

an exterior Maxwellian, this influence is ignored. Requiring a certain exterior
Maxwellian may produce solutions which have unphysical drops in the fluid
variables at the boundary. Methods for handing this problem are discussed
in the next chapter.

1.3   Radiation Transport

The starting point for a discussion of radiative heat transfer could be
Maxwell's equations for the propagation of electro-magnetic radiation. In
the limit of geometric optics, when the wavelength of the radiation is much
smaller than the macroscopic dimensions, the waves may be treated as photon particles which travel at the speed of light. In this case much of the
framework already developed for a system of gas molecules may be applied
to a system of photons. A Liouville equation may be written and integrated
to give a Boltzmann like equation for a one particle photon density function.

This analysis is useful for understanding the similarities between rarefied gas dynamics and radiative heat transfer, and in particular, the similarities in the numerical methods. On the other hand, the analogy is in
some ways artificial because of the very different nature of the particle systems. The photon system consists of non-interacting particles which are
subject to forces only in that they may scatter off boundaries as well as in
the interior of the domain. Moreover, particles may be absorbed or emitted
throughout the domain. The result is a linear equation with a scattering
operator which does not conserve particle number. This is opposed to the
nonlinear gas dynamics Boltzmann equation with its conservative collision
operator. Because the photons in the particle system act independently, the
assumption of molecular chaos for the two particle distribution function is
automatically satisfied, and thus it is only necessary to consider the one
particle distribution function. The resulting Boltzmann equation is known
as the radiation transport equation.

In the following we will focus on the radiation transport equation as
it is most commonly used - as a steady state equation for either radiance
or radiation intensity. These are quantities which have various, occasionally
conflicting, definitions in the literature. Therefore we will attempt to provide
clear definitions with units and physical interpretation. The bulk of the
analysis will be devoted to boundary conditions, which play the dominant
role in the radiation transport problems which arise in microelectronics
processing.

1.3.1 The Radiation Transport Equation

Radiation transport in the limit of geometrical optics is generally described by a linear Boltzmann equation of the form

$$\frac{\partial f}{\partial t} + c\,\omega\cdot\nabla_x f = \int K(x,\kappa \leftarrow \kappa')f(x,\kappa')d\kappa' - \nu(x,\kappa)f(x,\kappa) + \phi(x,\kappa) \quad (1.65)$$

The speed of all particles, the photons, is equal to the speed of light c. The direction of the particles is $\omega \in S^2$, the unit sphere. κ is the photon wavenumber vector, $\hbar\kappa$ is the photon momentum ($\hbar$ being Planck's constant), $\tilde\nu = |\kappa|$ is the wavenumber, and $\nu = c\,|\tilde\nu|$ is the frequency. The first term on the right hand side describes the gain of the number of photons with wavenumber κ by scattering from wavenumber κ' into κ. The second term describes the loss of photons with wavenumber κ due to absorption or scattering. The last term describes a volume radiation source.

Usually the radiation transport equation is written in terms of the radiance instead of the photon distribution function. These concepts are related as follows. The photon distribution function $f(x,\omega,\tilde\nu)$ gives the number of photons in the volume element $d^3x\ \tilde\nu^2 d\tilde\nu\ d\omega$ (polar coordinates in momentum space), where $d\omega$ is a solid angle element, as

$$dN = f(x,\omega,\tilde\nu)d^3x\ \tilde\nu^2 d\tilde\nu\ d\omega = f(x,\omega,\nu)\frac{1}{c^3}d^3x\ \nu^2 d\nu\ d\omega \quad (1.66)$$

The energy of these photons is, according to the Einstein relation $E = h\nu$

$$dE = f(x,\omega,\nu)\frac{h}{c^3}\nu^3 d^3x\ d\nu\ d\omega =: i_\nu(x,\omega)d^3x\ d\nu\ d\omega \quad (1.67)$$

$i_\nu(x,\omega)$, the radiance, is the energy density per unit solid angle per unit frequency interval and has the units

$$[i_\nu] = \frac{J}{m^3}\frac{1}{1/s}\frac{1}{sr} \quad (1.68)$$

From the radiance, the spectral radiation intensity, i.e. the radiation flux per unit solid angle per unit frequency interval, can be derived. This is the appropriate quantity for describing surface to surface radiation. Let $n(x)$ be the normal vector of a surface intersecting the radiation. The radiation moves with the speed of light c towards the wall. The spectral radiation intensity is then defined as

$$I_\nu(x,\omega) = c\,|n(x)\cdot\omega|\,i_\nu(x,\omega) \quad (1.69)$$

and has the units

$$[I_\nu] = \frac{Watt}{m^2}\frac{1}{1/s}\frac{1}{sr} \quad (1.70)$$

An example of the concepts defined above is isotropic black body radiation
with the Planck distribution function

$$f(x,\omega,\nu) = \frac{2}{e^{h\nu/k_B T(x)} - 1} \tag{1.71}$$

from the Bose-Einstein statistics. The factor of two takes the two possible
photon polarization states into account. The spectral radiation intensity is
then

$$i_{\nu,T}^{BB}(x) = \frac{2h}{c^3} \frac{\nu^3}{e^{h\nu/k_B T(x)} - 1} \tag{1.72}$$

The total energy leaving a surface is then given by the expression (the
Stefan-Boltzmann radiation law)

$$\int_0^\infty \int_{\Omega^+} I_{\nu,T}^{BB}(x,\omega)\, d\omega\, d\nu = \int_0^\infty \int_{\Omega^+} c\, i_{\nu,T}^{BB}(x)\, |n(x)\cdot\omega|\, d\omega\, d\nu = \sigma T(x)^4 \tag{1.73}$$

where $\sigma = 5.6697 \times 10^{-8}\, Wm^{-2}K^{-4}$ is the Stefan-Boltzmann constant.

The further discussion of the radiation transport equation will be based
on the following approximations. First, only the stationary case will be studied since the time scale of the radiation transport within distances of about
1 meter is very small for the applications in the context of microelectronic
processing. Next, the radiation will be considered as unpolarized only. Polarization can be taken into account here by considering a three component
transport equation similar to the multi-species equation for the transport
equation of molecules or the multi-band equation for the transport of electrons. Polarization effects are not important for the simulation of radiative
heating or pyrometric sensors. Furthermore, inelastic scattering effects of
photons will not be considered. This excludes mechanisms like fluorescence
and stimulated emission. In this approximation, there is no interaction of
photons with different frequency ν in Equation (1.65).

The linear Boltzmann equation in terms of the radiance reads

$$(\omega \cdot \nabla_x + \beta_{\nu,T})\, i_\nu(x,\omega) = \sigma_{\nu,T} \int_\Omega g_\nu(x,\omega \leftarrow \omega')i_\nu(x,\omega')d\omega' - \epsilon_{\nu,T}\, i_{\nu,T}^{BB}(x) \tag{1.74}$$

The loss consists of a scattering contribution with the total scattering
coefficient $\sigma_{\nu,T}(x)$ and a volume absorption contribution $\alpha_{\nu,T}(x)$ so that
$\beta_{\nu,T}(x) = \sigma_{\nu,T}(x) + \alpha_{\nu,T}(x)$. The gain due to scattering is described by
the integral term, where $\sigma_{\nu,T}(x)g_{\nu,T}(x,\omega \leftarrow \omega')$ is the differential scattering coefficient. The inhomogeneous term describes a volume radiation source
with a spectral distribution as the black body distribution times the volume
emission coefficient $\epsilon_{\nu,T}(x)$.

The above equation describes the transport of unpolarized radiation
in a homogeneous medium. This implies that the effects of refraction in a

medium with non-constant refractive index are not included. But the effect of refraction at a surface where the refractive index changes discontinuously can be included in boundary conditions.

The integro-differential equation (1.74) can be transformed into an integral equation. The advantage of such a formulation is that the boundary conditions present in the problem can be incorporated explicitly. But this transformation of the integro-differential equation into an integral equation is only possible when the equation has the linear structure of (1.65). When we consider the change of i_ν in the direction ω only, we can integrate along the path as long as we do not hit a boundary, and we obtain [29]

$$
\begin{aligned}
i_\nu(x,\omega) = & \qquad\qquad\qquad\qquad\qquad\qquad\qquad\qquad\qquad (1.75)\\
& e^{-\int_0^s \beta_{\nu,T}(x-s'\omega)ds'} i_\nu(x-s\omega,\omega)\\
& + \int_0^s e^{-\int_0^{\tilde{s}} \beta_{\nu,T}(x-s'\omega)ds'} \epsilon_{\nu,T}(x)\, i_{\nu,T}^{BB}(x-\tilde{s}\omega)d\tilde{s}\\
& + \int_0^s e^{-\int_0^{\tilde{s}} \beta_{\nu,T}(x-s'\omega)ds'} \sigma_{\nu,T}(x) \int_\Omega g_\nu(x-\tilde{s}\omega, \omega \leftarrow \omega')i_\nu(x-\tilde{s}\omega,\omega')d\omega' d\tilde{s}.
\end{aligned}
$$

This can be proved by taking the derivative with respect to s on both sides. Inside a bounded domain, for every direction ω there is a parameter s such that $x - s\omega =: y$ lies on a boundary.

In the following applications, there will be the further simplification that volume scattering will be neglected, i.e. $\sigma_{\nu,T}(x) = 0$. This scattering effect, which is most important for atmospheric optics applications, is of no importance in the simulation of radiative heat transfer in microelectronic processing equipment. This leaves from the linear Boltzmann equation in (1.65) with only free photon transport and the volume sources and sinks.

Under these assumptions, questions of interest for the gas dynamics Boltzmann equation, such as relaxation to equilibrium and hydrodynamic limit equations, no longer arise. Instead it is the boundary conditions given by the geometry and nature of the processing equipment which completely determine the particle distribution function. These boundary conditions may be quite complicated, and thus we now turn our attention to them.

1.3.2 Boundary Conditions

At a boundary to an opaque solid, the radiation can only be reflected or absorbed, but not transmitted. The relation between the incident radiance $i_\nu^{(i)}$ and the reflected radiance $i_\nu^{(r)}$ is given by the bidirectional reflectivity function $\rho_{\nu,T}(x,\omega \leftarrow \omega')$ which fulfills reciprocity $\rho_{\nu,T}(x,-\omega' \leftarrow -\omega) =$

$\rho_{\nu,T}(x, \omega \leftarrow \omega')$ and the defining relation similar to (1.58)

$$|n(x) \cdot \omega| \, i_\nu^{(r)}(x, \omega) d\omega = \int_{\Omega^-} \rho_{\nu,T}(x, \omega \leftarrow \omega') |n(x) \cdot \omega'| \, i_\nu^{(i)}(x, \omega') d\omega', \ \omega \in \Omega^+$$
(1.76)

where the radiation is incident within a solid angle $d\omega$ from the half space Ω^- with $n(x) \cdot \omega < 0$ and is reflected within the solid angle $d\omega'$ in the half space Ω^+. There are other conventions for defining the bidirectional reflectivity. The convention mostly used in engineering radiative heat transfer is [87]

$$\rho_{\nu,T}^{siegel}(x, \omega \leftarrow \omega') = \frac{1}{|n(x) \cdot \omega|} \, \rho_{\nu,T}(x, \omega \leftarrow \omega')$$
(1.77)

The reflectivity furthermore has to fulfill

$$\int_{\Omega^+} \rho_{\nu,T}(x, \omega \leftarrow \omega') d\omega \leq 1$$
(1.78)

analogous to (1.60). Typical is specular reflection with

$$\rho_{\nu,T}^{spec}(x, \omega \leftarrow \omega') = r_{\nu,T}(x, \omega') \delta(\omega - \mathcal{R}(\omega'))$$
(1.79)

with the reflection operator $\mathcal{R}(\omega') = \omega' - 2\omega'(n(x) \cdot \omega')$ and $r_{\nu,T}(x, \omega')$ given by the Fresnel formulas (5.2) which correspond to electro-magnetic waves which are reflected at plane surfaces. The specular reflected radiance and radiation intensity are

$$
\begin{aligned}
i_\nu^{(r,spec)}(x, \omega) &= r_{\nu,T}(x, \omega') \, i_\nu^{(i)}(x, \omega') \\
I_\nu^{(r,spec)}(x, \omega) &= r_{\nu,T}(x, \omega') \, I_\nu^{(i)}(x, \omega')
\end{aligned}
$$
(1.80)

Another typical case is diffuse reflection with

$$\rho_{\nu,T}^{diff}(x, \omega \leftarrow \omega') = r_{\nu,T}(x, \omega') \frac{|n(x) \cdot \omega|}{\pi}$$
(1.81)

which corresponds to a maximal scattering from the roughness of the surface. The diffuse reflected radiance and radiation intensity are

$$
\begin{aligned}
i_\nu^{(r,diff)}(x, \omega) &= \int_{\Omega^-} r_{\nu,T}(x, \omega') \frac{|n(x) \cdot \omega'|}{\pi} \, i_\nu^{(i)}(x, \omega') d\omega' \quad (1.82) \\
I_\nu^{(r,diff)}(x, \omega) &= \frac{|n(x) \cdot \omega|}{\pi} \int_{\Omega^-} r_{\nu,T}(x, \omega') \, I_\nu^{(i)}(x, \omega') d\omega'
\end{aligned}
$$

The remaining fraction of radiation $1 - r_{\nu,T}(x, \omega')$ is transmitted through the surface. Because the absorption coefficient of opaque solids are very large, all the transmitted radiation is absorbed within a small distance of the surface - within much less than a micron for the case of metals. In the situation when such a small scale must not be resolved, the absorption is assumed to take place in the surface itself and the directional absorptivity is defined as $a_{\nu,T}(x, \omega') = 1 - r_{\nu,T}(x, \omega')$.

Another phenomenon is the emission of radiation from within a solid boundary. Let x be the position of the boundary and let $\epsilon_{\nu,T}(x - s\omega)\, i_{\nu,T}^{BB}(x - s\omega)$ be an isotropic volume radiation source in the depth of a homogeneous medium with $\epsilon_{\nu,T}$ the volume emissivity. The emitted radiation is partially absorbed on the path to the boundary, where $\alpha_{\nu,T}$ is the volume absorptivity and

$$t_{\nu,T}(x, x - s\omega) := e^{-\int_0^s \alpha_{\nu,T}(x - s'\omega)ds'} = e^{-s\alpha_{\nu,T}} \tag{1.83}$$

is the transmission from $x - s\omega$ to x. The radiation arriving at the boundary is then from (1.75)

$$\int_0^s e^{-s\alpha_{\nu,T}}\epsilon_{\nu,T}\, i_{\nu,T}^{BB}\, d\tilde{s} = \left(1 - e^{-s\alpha_{\nu,T}}\right)\frac{\epsilon_{\nu,T}}{\alpha_{\nu,T}}\, i_{\nu,T}^{BB} = \left(1 - t_{\nu,T}(s,0)\right)\frac{\epsilon_{\nu,T}}{\alpha_{\nu,T}}\, i_{\nu,T}^{BB} \tag{1.84}$$

The generalized Kirchhoff law states that $\epsilon_{\nu,T}(x) = \alpha_{\nu,T}(x)$. The solid is considered as opaque when the thickness is large compared to the absorption length, i.e. $s\alpha_{\nu,T} \to \infty$.

$$\lim_{s\alpha_{\nu,T}\to\infty} \left(1 - e^{-s\alpha_{\nu,T}}\right)\frac{\epsilon_{\nu,T}}{\alpha_{\nu,T}}\, i_{\nu,T}^{BB} = \frac{\epsilon_{\nu,T}}{\alpha_{\nu,T}}\, i_{\nu,T}^{BB} = i_{\nu,T}^{BB} \tag{1.85}$$

From this radiation arriving at the surface from the inside, a fraction of $r_{\nu,T}(x, \omega)$ is reflected back and a fraction of $1 - r_{\nu,T}(x, \omega)$ is transmitted (and refracted) through the surface. This fraction is then defined as the emissivity of the surface, i.e. $e_{\nu,T}(x, \omega) = 1 - r_{\nu,T}(x, \omega)$. The intensity or radiance of the surface is then

$$i_{\nu}^{(e)}(x, \omega) = e_{\nu,T}(x, \omega)\, i_{\nu,T}^{BB}(x) \tag{1.86}$$

$$I_{\nu}^{(e)}(x, \omega) = e_{\nu,T}(x, \omega)\, I_{\nu,T}^{BB}(x, \omega) = e_{\nu,T}(x, \omega)\, c\, i_{\nu,T}^{BB}(x)|n(x) \cdot \omega|$$

Kirchhoff's law is now the equality between spectral absorptivity and emissivity

$$e_{\nu,T}(x, \omega) = a_{\nu,T}(x, \omega) \tag{1.87}$$

When the solid is semitransparent and not opaque, the transmitted fraction of radiation can propagate further into the medium. The relation between the radiation $i_{\nu}^{(i)}$ incident on the boundary from the solid side and the transmitted radiation $i_{\nu}^{(t)}$ is given by the bidirectional transmission function $\tau_{\nu,T}(x, \omega \leftarrow \omega'')$ and the relation

$$|n(x) \cdot \omega|\, i_{\nu}^{(t)}(x, \omega)d\omega = \int_{\Omega+} \tau_{\nu,T}(x, \omega \leftarrow \omega'')|n(x) \cdot \omega''|\, i_{\nu}^{(i)}(x, \omega'')d\omega'', \quad \omega \in \Omega^+ \tag{1.88}$$

where the incident radiation is integrated over the half space Ω^+ with $n(x) \cdot \omega'' > 0$. In many cases, the transmitted direction will be given by Snell's law, $\omega = \mathcal{S}(\omega'')$, with the transmission function

$$\tau_{\nu,T}(x, \omega \leftarrow \omega'') = \left(1 - r_{\nu,T}(x, \omega'')\right)\delta(\omega - \mathcal{S}(\omega'')). \qquad (1.89)$$

The boundary condition at an arbitrary solid or semitransparent surface is now

$$i_\nu^{(r,t)}(x, \omega) = \int_\Omega \frac{|n(x) \cdot \omega'|}{|n(x) \cdot \omega|} \, \varsigma_{\nu,T}(x, \omega \leftarrow \omega')i_\nu^{(i)}(x, \omega')d\omega' \ , \ \omega \in \Omega^+ \quad (1.90)$$

The incident radiance has a contribution from radiation transported from another surface and from radiation emitted in the volume. The outgoing radiance has a contribution from surface emission and from reflection and transmission. The generalized transmission-reflection function is

$$\varsigma_{\nu,T}(x, \omega \leftarrow \omega') = \left\{ \begin{array}{l} \rho_{\nu,T}(x, \omega \leftarrow \omega') \ , \ \omega' \in \Omega^- \\ \tau_{\nu,T}(x, \omega \leftarrow \omega') \ , \ \omega' \in \Omega^+ \end{array} \right. \qquad (1.91)$$

When the transport equation (1.75) is combined with the boundary condition, a self-consistent equation for the radiance leaving a surface is obtained

$$\begin{aligned} i_\nu(x, \omega) \ = \ & \int_\Omega \frac{|n(x) \cdot \omega'|}{|n(x) \cdot \omega|} \, \varsigma_{\nu,T}(x, \omega \leftarrow \omega')\Big(t_{\nu,T}(x, x - s\omega')\, i_\nu(x - s\omega', \omega') \\ & + \int_0^s t_{\nu,T}(x, x - \tilde{s}\omega')\epsilon_{\nu,T}(x)\, i_{\nu,T}^{BB}(x - \tilde{s}\omega')d\tilde{s} \Big)d\omega' \quad (1.92) \end{aligned}$$

s is the distance in the direction $-\omega'$ to the next boundary which is hit at the point $y := x - s\omega'$. More precisely, $s = s(\omega')$ and $y = y(\omega')$ are functions of ω'. $i_\nu(x - s\omega', \omega')$ is the radiance leaving the surface at y. The fraction $i_\nu^{(t)}(x, \omega)$ of the radiation transmitted through the surface is obtained by limiting the integral to the half space Ω^+. Similarly, the reflected fraction $i_\nu^{(r)}(x, \omega)$ is obtained by integrating only over Ω^-.

The radiation transport equation (1.92) can now be solved with the help of Monte Carlo methods. The treatment of semitransparent regions can be further simplified when these are plane parallel. This issue is addressed in greater detail in Chapter 5.

1.4   Electron Transport

We now consider the evolution of systems of charge carriers (specifically electrons - the transport of holes is discussed below). Initially we will consider a collection of charged classical particles (an electron gas) in a vacuum. From the Liouville equation for this system a Vlasov-Poisson equation will be derived in an analogous manner to the derivation of the Boltzmann equation for rarefied gases (Equation (1.16)). The effects of a background lattice of positive ions, which play a dominant role in electron transport in semiconductors, are then introduced to give the semi-classical Liouville and Vlasov-Poisson equations. Here effects due to the quantum nature of matter first appear. Finally the semi-classical Boltzmann equation is derived by considering scattering effects and short range particle-particle interactions (collisions). From this Boltzmann equation, as in Sections 2.2 - 2.3, various limiting equations and equilibrium may be deduced. A much more detailed account of this derivation procedure, upon which the following is based, can be found in [60].

1.4.1 Classical Vlasov-Poisson

As with the system of gas molecules, we begin by describing the electron gas in a vacuum by a probability density function $P^N(\vec{x}, \vec{p}, t)$ (see Equation (1.1)). Again the particles move according to Equations (1.2) resulting in the Liouville equation (Equation (1.6)). In the case of classical particles ($\vec{p} = m\vec{v}$), this may be written

$$\frac{\partial P^N}{\partial t} + \vec{v} \cdot \vec{\nabla}_{\vec{x}} P^N + \frac{1}{m} \vec{F} \cdot \vec{\nabla}_{\vec{v}} P^N = 0 \,. \tag{1.93}$$

Again Equation (1.7) is used to define the one particle probability density function, and Equation (1.93) is integrated to give and equation for P^1.

The difference between the rarefied gas case and the electron gas case is manifested in the force term $\vec{F}$. For the charge neutral gas molecules it was assumed that no external force fields were present and that the interparticle forces were extremely short range - on the order of a molecular diameter d. The Boltzmann equation was obtained in the limit $Nd^3 \to 0$ but $Nd^2 > 0$.

For the system of electrons the force $\vec{F}$ consists of an external field E_{ext} acting on the charged particles, as well as weak long range inter-particle forces. The force on particle i may then be expressed as

$$F_i = -qE_{ext}(x_i) - q \sum_{j=1, j \neq i}^{N} E_{int}(x_i, x_j) \,. \tag{1.94}$$

Here q is the elementary unit of charge such that an electron has charge $-q$. Typically a Coulomb force is used for E_{int} which becomes singular as two particles approach. Difficulties here are avoided by considering the limit $Nd^2 \to 0$, where d is now the effective interaction diameter of a screened electron. Thus the electron gas is in the free flow regime, and the resulting equation does not contain a collision term. Integrating Equation (1.93) and neglecting terms of order $\mathcal{O}(1/N)$ gives

$$\frac{\partial P^1}{\partial t} + v \cdot \nabla_x P^1 - \frac{q}{m} E_{ext} \cdot \nabla_v P^1 \tag{1.95}$$

$$- \frac{q}{m} \nabla_v \cdot \left(\int \int N E_{int}(x, x') P^2(x, v, x', v') \, dx' \, dv' \right) = 0 .$$

Here $E_{int}(x, x')$ must be $\mathcal{O}(1/N)$ so that $N E_{int}(x, x')$ remains finite. This is consistent with free flow regime assumption as well as with the idea that the total force exerted on an electron should remain finite as $N \to \infty$. Again the assumption of molecular chaos

$$P^2(x, v, x', v') = P^1(x, v) P^1(x', v') \tag{1.96}$$

is necessary to close Equation (1.95).

For electrons it is most convenient to work in terms of phase space number density and position space number density, defined as

$$F(x, v, t) = N P^1(x, v, t) \tag{1.97}$$

$$n(x, t) = \int F(x, v, t) \, dv .$$

F represents the (expected) number of particles to be found in a given state (x, v). The Pauli exclusion principle, one of the fundamental laws of quantum mechanics, states that at most one electron may be found in any given state. This law introduces the restriction that

$$0 \leq F(x, v, t) \leq 1 . \tag{1.98}$$

Using Equation (1.97) and assumption (1.96), Equation (1.95) may be rewritten as

$$\frac{\partial F}{\partial t} + v \cdot \nabla_x F - \frac{q}{m} E_{eff} \cdot \nabla_v F = 0 \tag{1.99}$$

where

$$E_{eff}(x, t) = E_{ext}(x, t) + \int n(x', t) E_{int}(x, x') \, dx' \tag{1.100}$$

is the effective electric field acting on a particle at position x. It may be shown that if the initial data for Equation (1.99) satisfy Inequality (1.98), then this condition remains true for all $t > 0$.

Equation (1.100) may be replaced by a differential equation, the Poisson equation, if it is assumed that the external field is the gradient of an external potential

$$E_{ext}(x, t) = -\nabla_x V_{ext}(x, t) , \tag{1.101}$$

and the inter-particle field is given through a Coulomb force

$$E_{int}(x, x') = -\frac{q}{4\pi\epsilon}\frac{x - x'}{|x - x'|^3} .$$ (1.102)

Because the Coulomb force is also a gradient, it follows that $E_{eff}(x, t)$ is the gradient of a potential

$$E_{eff}(x, t) = -\nabla_x V_{eff}(x, t) ,$$ (1.103)

By taking the divergence of Equation (1.100), the Poisson equation for V_{eff} is obtained:

$$\Delta V_{eff} = \Delta V_{ext} + \frac{q}{\epsilon}n .$$ (1.104)

Finally, if the external field is in fact due to a Coulomb force generated by a background of positively charged ions of charge $+q$ and with number density $C(x, t)$,

$$E_{ext}(x, t) = \frac{q}{4\pi\epsilon}\int C(x', t)\frac{x - x'}{|x - x'|^3} dx' ,$$ (1.105)

then Equation (1.104) becomes

$$\Delta V_{eff} = -\frac{q}{\epsilon}(C - n) .$$ (1.106)

The coupled system of Equations (1.99), (1.103) and (1.104) is known as the Vlasov-Poisson equation.

1.4.2 Semi-Classical Vlasov-Poisson

Next we turn our attention to the derivation of the semi-classical transport equations which are obtained by including certain quantum effects associated with the presence of the crystal lattice of positive ions. The electronic properties of semiconductor materials are primarily determined by these effects, so that their inclusion is fundamental to accurate equations for charge transport.

The fields of solid state physics and quantum mechanics are both extensive and extensively described in the literature. Therefore we give here only a superficial introduction, the intention of which is to motivate the use of a new basic independent variable k, the electron wave vector, to replace the particle velocity in the classical equations. The use of k instead of v represents the fundamental difference between the semi-classical and the classical transport equations.

It may be helpful to first consider the qualitative differences among conductors, insulators and semi-conductors. A basic property of matter is that the electrons in an atom are restricted to distinct energy levels, known as bands. At a temperature of absolute zero, all electrons would lie in the lowest energy valence bands. An electron in a valence band is tightly bound to the central nucleus and is not available to carry charge through a material. As the temperature increases, thermal fluctuations may allow some electrons to move to higher energy bands, known as conduction bands. Here the electrons are only weakly tied to the nucleus, and are therefore free to move through the material and conduct electricity. The energy difference between the highest valence band and the lowest conduction band is called the band gap. Conductors are materials where the Fermi energy, which describes the energy level of the occupied states, is within one band. Hence the valence band is not completely filled and there is in a sense a zero band gap. Insulators, on the other hand, have a filled valence band and large band gaps which effectively insure that no electrons will be in the conduction band. Semiconductors are materials (including Silicon (Si), Gallium-Arsenide (GaAs) and some less commonly used III-V compounds) which have a filled valence band and smaller band gaps, which may be easily modified by introducing impurities (doping) to become conductors. By combining two different doped materials, it is possible, for example, to produce a device which allows current to flow only in one direction (the pn-junction).

The essential properties of a semiconductor are determined by the energy bands and the relationship (called the band structure) between the energy and the momentum of an electron within a given band. Mathematically, those bands correspond to eigenstates of the Schrödinger equation for an electron moving in a periodic potential.

We consider now an electron moving through a lattice of positively charged ions which creates a lattice periodic potential $V_{\mathcal{L}}$. Let

$$\mathcal{L} = \{i\vec{a_1} + j\vec{a_2} + l\vec{a_3} : i, j, l \in Z\} \qquad (1.107)$$

be the crystal lattice defined by the base vectors $\vec{a_1}, \vec{a_2}, \vec{a_3} \in R^3$. Define the corresponding reciprocal lattice

$$\mathcal{R} = \{i\vec{b_1} + j\vec{b_2} + l\vec{b_3} : i, j, l \in Z\} \qquad (1.108)$$

by

$$\vec{a_i} \cdot \vec{b_j} = 2\pi\delta_{ij} \quad i, j = 1, 2, 3 \qquad (1.109)$$

with δ_{ij} the Kronecker delta function. A primitive cell of the lattice is defined as any translation of the parallelepiped determined by the base vectors. There is one primitive cell of the reciprocal $\mathcal{R}$ lattice which is of particular interest. Known as the Brillouin zone B, it is the primitive cell which is centered at the origin, which is to say that it contains all points in R^3 which are closer to the origin than to any other point of $\mathcal{R}$. B is necessarily symmetric, so that $k \in B$ if and only if $-k \in B$. As discussed below, an electron may be described by a function of x, periodic on $\mathcal{L}$, and of k, periodic on $\mathcal{R}$. The symmetry of B makes it convenient to take it as the domain for k.

From the theory of quantum mechanics, an electron in the lattice is described by a state function $\psi(x)$ which satisfies the Schrödinger equation

$$-\frac{\hbar^2}{2m}\Delta\psi - qV_{\mathcal{L}}\psi = \mathcal{E}\psi . \tag{1.110}$$

Here $\mathcal{E}$ is the energy of the electron and $\hbar$ is Planck's constant. Bloch's theorem states that every solution of Equation (1.110) may be expressed as a plane wave modulated by a lattice periodic function, i.e.

$$\psi(x) = e^{ik\cdot x}u_k(x) \tag{1.111}$$

with

$$u_k(x + X) = u_k(x) \quad X \in \mathcal{L} . \tag{1.112}$$

Here k is an arbitrary vector in $\mathbf{R}^3$. Moreover, there are such solutions for every k. When k is fixed, substitution of Equation (1.111) into Equation (1.110) leads to an elliptic eigenvalue problem which has a countable set of eigenvalue, eigenstate pairs $\mathcal{E}^j$, $u_k^j(x), j \in N$. The eigenvalue is necessarily a function of k. The integer variable j indexes the allowable energy bands, while the function $\mathcal{E}^j(k)$ gives the energy structure of each band.

The eigenvalues $\mathcal{E}^j$ and eigenstates ψ^j are in fact periodic functions of the wave vector k. This may be seen by considering two solutions ψ_k^j and ψ_{k+K}^j with $K \in \mathcal{R}$, the reciprocal lattice. Then

$$\begin{aligned}\psi_{k+K}^j &= e^{i(k+K)\cdot x}u_{k+K}^j(x) \tag{1.113}\\ &= e^{ik\cdot x}\left(e^{iK\cdot x}u_{k+K}^j(x)\right) .\end{aligned}$$

By Equation (1.109) the quantity in the parentheses is also lattice periodic on $\mathcal{L}$, and thus by the Bloch decomposition

$$u_k^j(x) = e^{iK\cdot x}u_{k+K}^j(x) . \tag{1.114}$$

It follows then that

$$\begin{aligned}\psi_{k+K}^j &= \psi_k^j \tag{1.115}\\ \mathcal{E}^j(k+K) &= \mathcal{E}^j(k) .\end{aligned}$$

Therefore it is possible to restrict the domain of k to the Brillouin zone B. The quantity $\mathcal{E}^j(k)$ is called the jth energy band of the crystal lattice. It is determined by the lattice potential $V_{\mathcal{L}}$. In practise these quantities are not known analytically, but must be measured in experiments.

The semi-classical approximation replaces the wave function $\psi_k^j(x)$ with a particle located in the jth energy band with position x and associated wave vector $k \in B$. This crystal electron is no longer a free particle, but a quasi-particle which contains information about the underlying lattice through the function $\mathcal{E}(k)$. It is possible to show that the effect of a force F, constant or periodic in k, applied to the crystal electron is described by the equation

$$\hbar \dot{k} = F \; . \tag{1.116}$$

Thus the quantity $p = \hbar k$ behaves like the momentum of the quasi-particle, and is therefore termed the crystal momentum. It is then appropriate to define the crystal electron velocity as

$$v(k) = \frac{1}{\hbar} \nabla_k \mathcal{E}(k) = \nabla_p \mathcal{E}(\frac{p}{\hbar}) \; . \tag{1.117}$$

For a collection of N crystal electrons with positions $\vec{x} = (x_1, \ldots, x_N)$ and wave vectors $\vec{k} = (k_1, \ldots, k_N)$ the semi-classical equations of motion are again given by Equation (1.2):

$$\dot{\vec{x}} = \vec{v} \tag{1.118}$$
$$\dot{\vec{p}} = \vec{F}$$

where now $\vec{v} = (v(k_1), \ldots, v(k_N))$ and $\vec{p} = (p(k_1), \ldots, p(k_N))$ with the functions $v(k)$ and $p(k)$ defined by Equations (1.117) and (1.116).

In the limit that the lattice spacing of $\mathcal{L}$ shrinks to zero so that the electron sees essentially a constant potential $V_{\mathcal{L}}$, the eigenvalue of the Schrödinger equation becomes

$$\mathcal{E}(k) = \mathcal{E}_{min} + \frac{\hbar^2}{2m}|k|^2 \tag{1.119}$$

where the Brillouin zone B has now been enlarged to $\mathbf{R}^3$. Here $\mathcal{E}_{min} = -qV_{\mathcal{L}}$ is the minimum band energy. This is known as the parabolic band approximation. For $V_{\mathcal{L}} = 0$, this corresponds to the energy band of a free electron in a vacuum. In this approximation, Equations (1.117) and (1.116) give

$$v = \frac{p}{m} \tag{1.120}$$

and the classical dynamics are recovered.

The parabolic band approximation is very frequently used in applications when either the true band structure is not known, or when it may be assumed that the electron energies all lie near the band minimum. This is often the case when the external fields are not strong. The approximation may be improved by replacing the electron rest mass m with an effective mass m_{eff} which more accurately reflects the true band behavior.

The semi-classical Liouville equation, which corresponds to the classical Equations (1.6) and (1.93), now reads

$$\frac{\partial P^N}{\partial t} + \vec{v}(\vec{k}) \cdot \vec{\nabla}_{\vec{x}} P^N + \frac{1}{\hbar} \vec{F} \cdot \vec{\nabla}_{\vec{k}} P^N = 0 \tag{1.121}$$

with

$$x_i \in \mathbf{R}^3 \text{ and } k_i \in B \text{ for } i = 1, \ldots, N .$$

The exact same integration procedure applied to the classical electron gas Liouville equation may now be repeated to obtain the semi-classical equivalent of Equation (1.99)

$$\frac{\partial F}{\partial t} + v(k) \cdot \nabla_x F - \frac{q}{\hbar} E_{eff} \cdot \nabla_k F = 0 \tag{1.122}$$

where

$$E_{eff}(x, t) = E_{ext}(x, t) + \int n(x', t) E_{int}(x, x') \, dx' . \tag{1.123}$$

Again the number density F must satisfy the Pauli exclusion principle

$$0 \leq F(x, k, t) \leq 1 , \tag{1.124}$$

but now the position space number density $n(x, t)$ is defined by

$$n(x, t) = \int_B F(x, k, t) \, dk . \tag{1.125}$$

As before, if the inter-particle field is generated by a Coulomb force, then Equation (1.123) may be replaced by the Poisson equation (1.104).

1.4.3   Semi-Classical Boltzmann Equation

Ideally one would now return to the semi-classical Liouville equation (1.121) for a system of crystal electrons and modify the force term to include short range forces corresponding to collisions of the electrons with themselves as well as with the background lattice. These interactions, however, are considerably more complicated than the spherical potential type collisions of classical gas molecules. Various technical problems arise if one tries to integrate and pass limits on the Liouville equation when these forces are included. Therefore the semi-classical Boltzmann equation for electron transport in a crystal lattice is usually obtained as a phenomenological model, derived in the spirit of Boltzmann's original formulation for rarefied gases.

The key idea is that particles instantaneously scatter in momentum space (k-space) from k to k' at a fixed spatial position x with a certain rate $R(x, k \to k', t)$. This may be compared with the scattering processes used in radiation transport. In radiation scattering, however, the rate R is assumed proportional to $F(k)$, the probability of finding a particle in state k. For electrons, the Pauli exclusion principle requires that R also be proportional to $1 - F(k')$, which is the probability that the end state k' is not occupied. In general we write

$$R(x, k \rightarrow k', t) = s(x, k \rightarrow k', t)F(k)(1 - F(k')) \tag{1.126}$$

where s is called the scattering rate. The collision term in the Boltzmann equation is obtained by integrating the rate of scattering from state k' into state k minus the rate of scattering out of state k into state k' over $k' \in B$:

$$Q(F) = \int_B [s(x, k' \rightarrow k, t)F(k')(1 - F(k)) - s(x, k \rightarrow k', t)F(k)(1 - F(k'))]\, dk'. \tag{1.127}$$

The semi-classical Boltzmann equation then reads

$$\frac{\partial F}{\partial t} + v(k) \cdot \nabla_x F - \frac{q}{\hbar} E_{eff} \cdot \nabla_k F = Q(F) \tag{1.128}$$

which may again be coupled with the Poisson equation to determine E_{eff}.

For the case of electron scattering off the background lattice, $s(x, k \rightarrow k', t)$ will be a function only of crystal properties such as $\mathcal{E}(k)$. This is similar to the scattering rate of photons in radiation transport. For electron-electron scattering, however, the scattering rate must also depend on the local electron density. For a rarefied gas, the probability that a molecule of velocity v collides with a molecule of velocity v_* was proportional to $f(v)f(v_*)$. For a system of electrons, the Pauli exclusion principle must again be taken into account. The probability that an electron with wave vector k collides with an electron with wave vector k_* such that they scatter into states k' and k'_* is proportional to

$$F F_* (1 - F')(1 - F'_*)\delta_{\mathcal{E}} \delta_k . \tag{1.129}$$

Here $F = F(k)$, $F_* = F(k_*)$, $F' = F(k')$ and $F'_* = F(k'_*)$. The deltas are now Dirac delta functions representing conservation of energy and momentum

$$\begin{aligned} \delta_{\mathcal{E}} &= \delta(\mathcal{E}(k) + \mathcal{E}(k_*) - \mathcal{E}(k') - \mathcal{E}(k'_*)) \tag{1.130}\\ \delta_k &= \delta(k + k_* - k' - k'_*) \end{aligned}$$

The scattering rate $s(x, k \rightarrow k', t)$ for an electron-electron collision may be expressed as an integral over all possible collision partners k_*

$$s_{e-e} = \int_B \int_B \mathcal{B}(|k - k_*|, |k' - k'_*|)F_*(1 - F'_*)\delta_{\mathcal{E}} \delta_k \, dk_* dk'_* . \tag{1.131}$$

Here $\mathcal{B}$ plays the role of the collision cross section (introduced in Section 1.2). It is only a function of the relative momentum of the colliding pair. When electron-electron scattering is included, the collision term contains a fourth order nonlinearity.

1.4.4   Equilibrium

In Section 1.2.1. the relaxation to equilibrium and the equilibrium distribution function were discussed for the Boltzmann equation of rarefied gas dynamics. The Maxwellian equilibrium was obtained as a consequence of the fact that the collision operator conserves mass, momentum and energy. For electrons in a lattice, however, the most significant scattering processes involve collisions with the background so that neither momentum nor energy are conserved. (Conservation of electron number (or equivalently, charge) is guaranteed by the form of Equation (1.127)). Nonetheless, it is possible to prove [80] that when certain assumptions on the scattering rate s are satisfied, the equilibrium density distribution which makes $Q(F^{eq}) = 0$ is in fact a Fermi-Dirac distribution

$$F^{eq}(k) = \frac{1}{1 + e^r} \tag{1.132}$$

$$r(k) = \frac{\mathcal{E}(k) - \mathcal{E}_F}{k_B T} .$$

The constant $\mathcal{E}_F$ is called the Fermi energy, k_B is the Boltzmann constant, and T is the lattice temperature.

The necessary assumption on the scattering rate is that $s(k \to k')$ be a bounded measurable function which satisfies

$$M(k)\, s(k' \to k) = M(k')\, s(k \to k') . \tag{1.133}$$

Here $M(k)$ is a normalized Maxwellian distribution

$$M(k) = c\, e^{-r(k)} \tag{1.134}$$

$$c = \int_B e^{-r(k)} dk$$

for $r(k)$ given in Equation (1.132). Several inequalities of the form

$$\int_B Q(F) \chi \left[\frac{1 - F}{F} M \right] dk \geq 0 , \tag{1.135}$$

where $\chi(\cdot)$ is any increasing function, are also established in [80]. These may be compared with Equation (1.36).

The principle of detailed balance states that at equilibrium, the rates of forward and backward scattering between two states are equal, i.e.

$$s(k \to k') F(1 - F') = s(k' \to k) F'(1 - F) . \tag{1.136}$$

If F is assumed to be a Fermi-Dirac distribution as in Equation (1.132), then it follows easily that $s(k \rightarrow k')$ satisfies Equation (1.133). Thus condition (1.133) is necessary and sufficient for a Fermi-Dirac equilibrium.

Condition (1.133) naturally leads to the definition of a function $\phi(k, k')$:

$$\phi(k, k') = \frac{s(k \rightarrow k')}{M(k')} \qquad (1.137)$$

which is symmetric, i.e.

$$\phi(k, k') = \phi(k', k) . \qquad (1.138)$$

This function is known as the collision cross section. The collision operator may then be expressed as

$$Q(F) = \int_B \phi(k, k')[M(k)F'(1 - F) - M(k')F(1 - F')]dk' . \qquad (1.139)$$

Very often the low density approximation is made, which states

$$0 \leq F_{LD} \ll 1 . \qquad (1.140)$$

The collision term then becomes

$$Q_{LD}(F) = \int_B \phi(k, k')[M(k)F'_{LD} - M(k')F_{LD}]dk' . \qquad (1.141)$$

It is easily seen that condition (1.133) is equivalent to

$$F_{LD}^{eq}(k) = M(k) \qquad (1.142)$$

in the low density approximation.

1.4.5   Hydrodynamic Equations

As mentioned above, only the number of electrons (or charge) is conserved by the Boltzmann collision operator. Thus there is only one intrinsic conservation law which may be obtained by integrating the semi-classical Boltzmann equation, as opposed to the 5 equations corresponding to conservation of mass, momentum and energy in the gas dynamics case (see Equation (1.42)). The conservation of charge equation is

$$-q\frac{\partial n(x, t)}{\partial t} + \nabla_x \cdot J(x, t) = 0 \qquad (1.143)$$

where

$$n(x, t) = \int_B F(x, k, t)dk \qquad (1.144)$$

is the electron number density and

$$J(x,t) = -q \int_B v(k)F(x,k,t)dk \tag{1.145}$$

is the electron current density.

A closed set of fluid dynamic equations is then obtained by finding a second equation for J and n. The most widely used choice for the additional equation is the approximation

$$J = \mu q n E_{eff} + q D_e \nabla_x n \tag{1.146}$$

where D_e is the electron diffusion coefficient and μ the electron mobility. The mobility is related to the diffusion coefficient with the Einstein relation

$$\mu = \frac{|q|D_e}{k_B T} = \frac{D_e}{|U_T|} \tag{1.147}$$

where $U_T = k_B T/q$ is the thermal voltage. It can be seen that $\nabla \cdot J$ is composed of a diffusion term in n (with diffusivity μU_T) and a convection or drift term. Equations (1.143) and (1.146) together are known as the drift diffusion equations. The system obtained by coupling the Poisson equation (1.104) for a self consistent electric field to Equations (1.143) and (1.146) is called the basic semiconductor device equations.

The validity of Equation (1.146) must be considered. Effectively all details of the scattering process have been ignored, so one can only expect to describe the physics accurately if the macroscopic gradients occur over length scales considerably larger than the collision length scale. This is the small Knudsen number regime (see Section 1.2.2). Small (sub-micron) devices or devices operating under high electric fields generally do not satisfy this assumption.

As described in [60], Poupaud has carried out a rigorous Hilbert expansion (see Equation (1.47)) of the non-dimensional semi-conductor Boltzmann equation in the low density approximation assuming a parabolic band structure. The small expansion parameter is the Knudsen number. As in the gas dynamics case, the leading order term is a Maxwellian

$$F^0(x,k,t) = n(x,t)M(k) \tag{1.148}$$

with $M(k)$ as in Equation (1.134). If Equation (1.146) is taken as a definition for J, then Equation (1.143) results as a solvability condition for the second term of the expansion. Therefore the drift diffusion equations may be derived rigorously under the above assumptions.

Another approach to deriving fluid type equations is the moment closure method. Here the Boltzmann equation is integrated against various velocity "moments" - typically $(1, v_x(k), v_y(k), v_z(k), |v(k)|^2)$. Equations similar to (1.42) will result, but the new equations will include force terms.

More importantly, however, the integral of the collision term is no longer zero for any but the first of these velocity moments. The system of equations is closed by assuming a functional form for the distribution function F which depends on 5 parameters (one for each moment equation). The most common choice is a shifted Maxwellian as in Equation (1.28) (as opposed to that of Equation (1.134)):

$$F(x, k, t) = c \, n \, \exp\left(\frac{-(v(k) - v_0)^2}{2RT_e}\right) . \tag{1.149}$$

Here $n(x, t)$, $v_0(x, t)$ and $T_e(x, t)$ are the five parameters, $c(v_0, T_e)$ is a normalizing factor so that F integrates over k-space to give n, and $R = k_B/m$ is the gas constant. T_e is called the electron temperature, which may be different than the lattice temperature.

Usually the parabolic band approximation (Equation (1.119)) is made. Then the classical equations may be used with $v \in R^3$ replacing $k \in B$ as the independent variable. The resulting fluid equations are called hydrodynamic semiconductor equations

$$\frac{\partial n}{\partial t} + \nabla \cdot (n v_0) = 0 \tag{1.150}$$

$$\frac{\partial v_0}{\partial t} + v_0 \cdot \nabla v_0 + \frac{1}{n}\nabla P + \frac{q}{m}E_{eff} = C_v$$

$$\frac{3}{2}n\frac{k_B}{m}\left(\frac{\partial T_e}{\partial t} + v_0 \cdot \nabla T_e\right) + P\nabla \cdot v_0 = C_T$$

where P is the electron pressure

$$P = n \, \frac{k_B}{m} \, T_e \tag{1.151}$$

and

$$C_v = \frac{1}{n}\int v \, Q(F) \, dv \tag{1.152}$$

$$C_T = \int \left(\frac{1}{2}v^2 - v_0 \cdot v\right) Q(F) \, dv .$$

If the collision terms are set to zero, Equations (1.150) reduce to the Euler equations for a fluid acting under the field E_{eff}. In general it is quite difficult to evaluate C_v and C_T, so that various approximations must be made to make the equations tractable.

In Chapter 8 two analytic models for $F(x, k, t)$ are considered. The models are compared with the Monte Carlo solution of the Boltzmann equation for the high energy tail of the distribution.

1.4.6 Electrons and Holes

When an electron leaves the valence band for the conduction band, it leaves a vacancy, or hole, in the valence shell. This hole will tend to be filled by neighboring valence electrons, thereby creating a new hole. In this way holes are said to move through the lattice in the valence band, just as electrons move through the conduction band. Because a hole represents the absence of an electron, it is viewed as a charge carrier with positive charge $+q$. A Boltzmann equation for holes may be written which is identical to Equation (1.128) with the exception that the sign of the charge is switched.

The significance of hole transport becomes apparent in light of the practise of semiconductor doping. A pure silicon crystal is by nature an insulator. The four electrons of the outer valence band bind with four other electrons, one from each of the four neighboring lattice atoms, to fill out the valence shell. The energy required to move one of these electrons to the conduction band (i.e., the band gap) is rather large. However, if impurities are introduced into the crystal which are either rich or deficient in valence electrons (compared to silicon), the situation changes. For example, if a phosphorus atom with five valence electrons is included in the crystal, four of the electrons will bind into the lattice. The fifth will have greater freedom and may more easily move into the conduction band. Likewise, if a boron atom, with only three valence electrons, is implanted in the crystal, there will be holes in the valence band. The process of introducing impurities is known as doping. A semiconductor doped with excess electrons is called n-type (for negative), while a hole rich semiconductor is called p-type.

A semiconductor device usually contains both p and n-type regions. It becomes necessary to solve the two species Boltzmann equation (see Equation (1.53)) for the charge carrier distribution functions for electrons and holes. The two equations of type (1.128) are combined through the inclusion of an extra term describing electron-hole recombination and generation. A recombination-generation rate similar to the scattering rate must be given. A condition on this rate of the form of condition (1.133) is often imposed. In the small mean free path limit, a coupled set of drift diffusion equations may again be derived [80].

1.4.7 Boundary Conditions

As with gas dynamics and radiation, it is necessary in the electron case to specify the electron distribution function as a boundary condition. At a point x on the domain boundary ∂D with inward normal $n(x)$ a value must be given for $F(x, k, t)$ for all k such that $v(k) \cdot n(x) > 0$.

Semiconductor devices are typically bounded by insulated regions, which are treated as reflecting walls for the electrons, and contact segments, which are inflow/outflow boundaries. Specular reflection (see Equation (1.54)) is generally used for the insulated regions, although if more detailed knowledge of the boundary interaction were available, the more general formulation of Section 1.2.5 could be used. As with gas dynamics, the inflow boundary condition at the contact segments is taken to be a flux weighted Maxwellian distribution as in Equation (1.63).

When the Boltzmann equation is coupled to the Poisson equation, it is also necessary to impose boundary conditions on either V_{eff} or E_{eff}. On the insulated segments, $E_{eff} \cdot n(x)$ is specified (Neumann conditions), while at the contact segments, through which a voltage bias is applied to the device, the bias V_b is specified for V_{eff} (Dirichlet conditions).

1.5   Summary

The introduction given here was intended to highlight the similarities, as well as the differences, among three areas of transport which may be described by a Boltzmann equation. Many details, both mathematical and physical, have been left out. In particular, effects due to the full quantum nature of electrons were not discussed in Section 4. As semiconductor devices become smaller, it is likely that such effects will play an increasing role. Modeling of transport which includes these effects is a field of current research.

In the following chapters we present several applied problems in microelectronics, together with discussion of some of the Monte Carlo methods used to solve them numerically. In addition to describing some new results in the field, we hope to illustrate how Monte Carlo methods are applied, when they are necessary, and how Monte Carlo solutions to the Boltzmann equation may be used to establish models which may later be used in less computationally expensive numerical methods.

Chapter 2

Modeling of Gas Flow

Gas flow is an essential mechanism of transporting reactants and materials to the surface of a wafer in microelectronic processes. Accurate modeling of these flows is therefore vital in a simulation to determine such properties as uniformity of deposition. In this chapter we present an overview of such flows and the equations and boundary conditions which describe them.

We begin with a brief look at some of the manufacturing equipment commonly in use, focusing on the nature of the gas flows which arise. Most reactors currently in use operate in the hydrodynamic or near hydrodynamic regime. Therefore we continue with a discussion of the continuum balance equations for a multiple species gas. The consideration of relaxation time scales and slip boundary conditions allows the continuum equations to be extended into the near hydrodynamic range. Next we address certain aspects of modeling true transition regime reactor flows. Finally, free molecular flow is considered, including a derivation of the Knudsen diffusion coefficient for a simple geometry.

Figure 2.1 provides an overview of the relevant processes with regard to the flow regime in which they occur. This takes into account both geometric scales $\mathcal{L}$ and the mean free path λ (mfp) of the gas. These results are presented for a typical reactor environment of argon at $800°K$. This allows pressure, given in units of *Pascal* and *Torr*, to be related to mfp. For other gases and temperatures the mfp may be larger or smaller. A reasonable range is given by the mfp of hydrogen H_2 at $1100°K$ which is 2.16 times larger and the mfp of tungsten hexafluoride (WF_6) at $700°K$ which is 0.404 times smaller. The top scale of Figure 2.1 also gives the diluteness δ/d as the ratio of the mean molecular spacing $\delta = n^{-1/3}$ to the molecular diameter d. The dilute gas assumption requires $\delta/d >> 1$, which is well fulfilled for the given ranges. This figure also shows the flow regimes in terms of Knudsen number $Kn = \lambda/\mathcal{L}$.

49

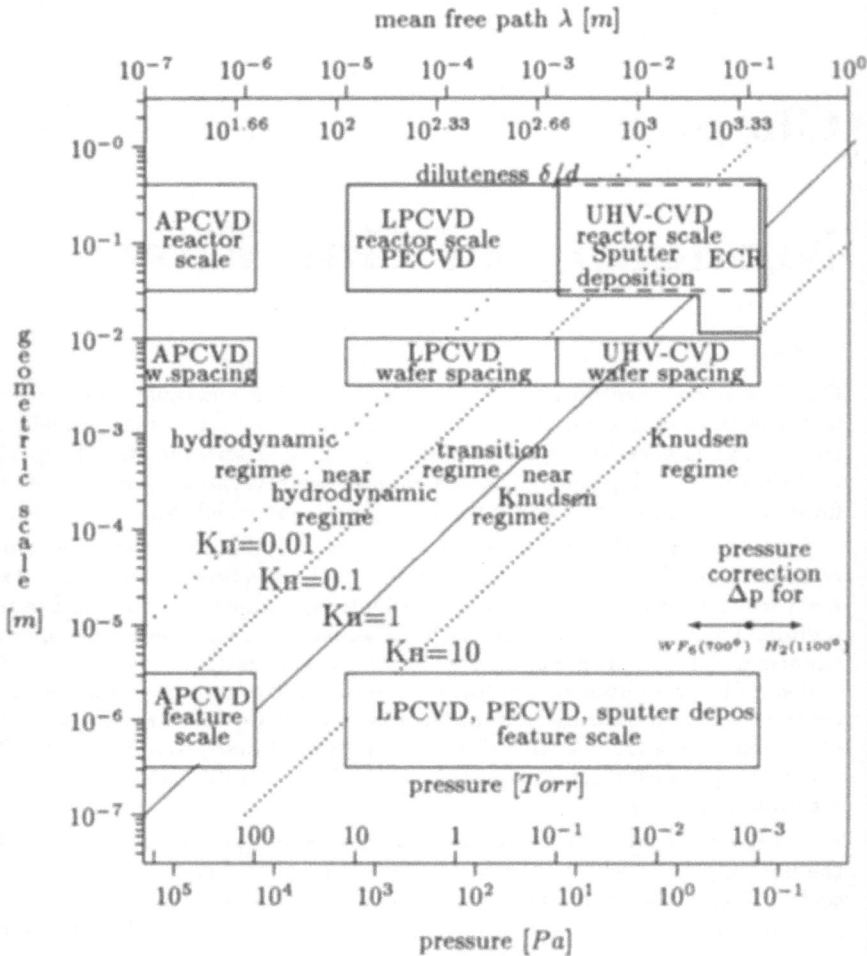

Figure 2.1: Deposition techniques in relation to the geometrical scale, the mean free path (mfp) and the relevant flow regimes. The vertical axis gives the geometrical scale. The top horizontal scale gives the mfp λ and the diluteness δ/d with reference to $d_{Ar} = 3.542 \times 10^{-10}$ m. The bottom horizontal scale gives the translation of the mfp into pressure in the units Pascal and Torr for the case of Ar at $800°K$. The shift of the pressure scale for other atmospheres can be estimated from the indicated pressure correction for WF_6 at $700°K$ and H_2 at $1100°$.

2.1   Typical Reactors

The basic principle of chemical vapor deposition (CVD) is that a suitable combination of reactant gases is brought in contact with a wafer surface that is maintained at an elevated temperature. One or more of the gas phase species reacts heterogeneously at the surface to deposit a solid film. Usually the activation energy is provided by the thermal energy of the wafer. In case of plasma enhanced chemical vapor deposition (PECVD) the activation energy is provided by the electrical energy of an ionized species. This allows the wafer to be held at a somewhat lower temperature. In the case of etching, the heterogeneous chemical reaction produces a volatile product which removes surface material.

CVD has become the primary technique to deposit thin films because of the good spatial control of film growth over a large surface area. Low Pressure Chemical Vapor Deposition (LPCVD) in horizontal or vertical multiple wafer reactors is the most important technique for depositing polycristalline Si and dielectric films such as SiO_2 and Si_3N_4 [50], [36]. The operating pressures range from a fraction of a $Torr$ to a few $Torr$. The deposition uniformity is very good since the Dahmköhler number, the ratio of the characteristic time for diffusion to the surface to the characteristic time for the surface reaction, is usually small in these cases.

The flow in this pressure range is in the hydrodynamic regime. In the interwafer regions, the flow might be in the near-hydrodynamic regime. Because the gradients in the interwafer regions are usually small, the corrections by slip boundary conditions or flux corrections (for example in the form of modified diffusion coefficients) will be small. Modeling and simulation in this regime have been widely studied [50], [49] and have reached a state where - provided a sufficient chemical model is given - it is possible to perform predictive simulation for design and optimization of equipment and processes for industrial CVD applications.

Recently, horizontal multiple wafer reactors have been used for the deposition of epitaxial silicon at a relatively low temperature in the $mTorr$ operating range [64]. This technique has been referred to as Ultra High Vacuum CVD (UHV-CVD). Although the growth rate is reduced at lower temperatures time so that longer processing times are required, the thermal budget is in the end reduced. This is because the activation energy for the diffusion falls off faster with temperature than the reaction rate for deposition, which is SiH_4 pyrolysis. The flow conditions at this pressure range from the near hydrodynamic regime to the transition regime to the near Knudsen regime in the interwafer region. The simulation of such processes requires methods for solving the Boltzmann equation. Such simulations were first done by Coronell et.al. [23], [24].

Physical vapor deposition (PVD) is a technique used for deposition of materials when suitable chemical reaction mechanisms are not known or are otherwise disadvantageous. In this technique, the film material is either evaporated or - in case of sputter deposition - removed from a source material with high energy ion beams. The evaporated or sputtered particles are then transported to the wafer surface where they condense to form the desired film. The deposited film must fulfil a number of requirements concerning uniformity, step coverage, stress, impurity concentration, resistance, morphology etc.

In all of the techniques mentioned above, the transport of the species or material to the wafer surface is just one factor, along with surface temperature and surface condition, which determines the characteristics of the deposited film. The solution of the transport problem provides the spatial densities of the species at the wafer surface including their energy distribution, angular distribution and the internal state.

2.2 Hydrodynamic Equations

We now consider the equations of motion for a continuum fluid. These may be derived by computing velocity moments of the Boltzmann equation, as was done in Section 1.2.2, or by applying physical arguments based on the conservation of mass, momentum and energy. The results of this are known as the balance equations, which express that the net change of a quantity in a fluid volume element with time is equal to the net flux of the quantity.

In order to obtain a closed system of equations, it is necessary to express the mass, momentum and energy fluxes in terms of the densities. In kinetic theory, this is known as the closure problem. One approach is to assume a functional form for the velocity distribution function $f(x, v, t)$ which includes the densities and possibly their derivatives as parameters. The moment equations then become a closed system for these parameters (assuming enough moments are taken). One difficulty of this approach is how to handle moments of the Boltzmann collision term. The validity of this method depends on how accurately the chosen functional form models the true distribution function. A new approach to moment closure has recently been studied in [57] and [58].

Another method for obtaining a closed system was already discussed in Section 1.2.2. Here the distribution function f is expanded asymptotically in terms of a small parameter ϵ corresponding to the Knudsen number. Truncation of the series leads to the desired closure. The Chapman-Engskog expansion leads to the Navier-Stokes equations which are discussed below. These equations are known to be valid for small ϵ (the hydrodynamic regime); it is difficult to predict how accurate they remain as ϵ increases.

A third approach is based on the Onsager theory of irreversible thermodynamics. This is not based on kinetic theory or the Boltzmann equation, although the resulting equations are qualitatively the same. The idea here is that the fluxes may be expressed as linear combinations of so-called generalized forces acting on the fluid. The appropriate generalized forces are quelled from an entropy balance equation. The use of thermodynamics dictates that the results will only be valid for flows close to thermal equilibrium. We summarize this approach below.

2.2.1 Balance Equations

In Section 1.2.2 the equations for a single component gas were given. We now extend this to a multiple species, chemically reacting flow and include external forces. The direct integration of the first velocity moments of the N species Boltzmann equation (Section 1.2.3) leads to equations for $\rho_i = m_i n_i$, $\rho_i \vec{v}_i$ and e_i, the species mass, momentum and internal energy densities. The total flow densities are then defined as

$$
\begin{aligned}
\rho &= \sum_{i=1}^{N} \rho_i \\
\rho \vec{v} &= \sum_{i=1}^{N} \rho_i \vec{v}_i \\
e &= \sum_{i=1}^{N} e_i \, .
\end{aligned}
\tag{2.1}
$$

The quantity $\vec{v}$ is called the mass averaged or stream velocity.

When thermodynamic relationships are included, it is often more convenient to work in terms of V, $\vec{v}$ and $\hat{u}$, the specific volume, specific momentum and specific internal energy. (A "specific" quantity means the quantity per unit mass, as opposed to a density which is the quantity per unit volume). The specific volume is defined as the reciprocal of the density: $V = 1/\rho$. The specific momentum is simply the stream velocity $\vec{v}$. The specific internal energy satisfies $\hat{u} = e/\rho$. The balance equations for the specific flow quantities may then be expressed with the help of the convective derivative $D/Dt = \partial/\partial t + \vec{v} \cdot \nabla$ as

$$
\rho \frac{DV}{Dt} = \nabla \cdot \vec{v}
\tag{2.2}
$$

$$
\rho \frac{D\vec{v}}{Dt} = -\nabla \cdot \underline{\underline{P}} + \sum_{i=1}^{N} n_i \vec{F}_i
\tag{2.3}
$$

$$
\rho \frac{D\hat{u}}{Dt} = -\nabla \cdot \vec{q} - \underline{\underline{P}} : \nabla \vec{v} + \sum_{i=1}^{N} n_i (\vec{v}_i - \vec{v}) \cdot \vec{F}_i \, .
\tag{2.4}
$$

Here $\underline{\underline{P}}$ is the pressure tensor representing the momentum flux, $\vec{q}$ is the heat flux vector, and $\vec{F}_i$ is the force acting on species i. The above system is completed by replacing the total mass balance equation (2.2) with the N individual species mass balance equations

$$\rho \frac{D\omega_i}{Dt} = -\nabla \cdot \vec{j}_i + \mathcal{N}_i \quad i = 1, \ldots, N . \tag{2.5}$$

Here $\omega_i = \rho_i/\rho$ is the mass fraction of species i. The diffusive mass flux vector $\vec{j}_i$ for species i is defined as

$$\vec{j}_i = \rho_i(\vec{v}_i - \vec{v}) . \tag{2.6}$$

The mass generation density of species i due to K chemical reactions with stochiometric coefficients ν_{ik} and forward and backward reaction rates $\mathcal{R}_k^g$ and $\mathcal{R}_{-k}^g$ (given in units $molecules/(m^3 s)$) is

$$\mathcal{N}_i(\vec{x}, t) = m_i \sum_{k=1}^{K} \nu_{ik}(\mathcal{R}_k^g - \mathcal{R}_{-k}^g) \tag{2.7}$$

The system of equations (2.3),(2.4) and (2.5) constitutes the equations of motion for a multiple species gas. If the mass, momentum and energy fluxes $\vec{j}_i$, $\underline{\underline{P}}$ and $\vec{q}$ can be expressed as functions of ρ_i, $\vec{v}$ and $\hat{u}$, then this becomes a closed system. This closure is addressed in the next section.

Before considering the closure, however, we first give an equation for the fluid temperature which can replace the internal energy equation (2.4). It is ofter more convenient to work with temperature than with internal energy. We begin by decomposing the pressure tensor $\underline{\underline{P}}$ into a static pressure p and a viscous stress tensor $\underline{\underline{\tau}}$

$$\underline{\underline{P}} = p\underline{\underline{I}} - \underline{\underline{\tau}} . \tag{2.8}$$

Here $\underline{\underline{I}}$ is the identity tensor. The static pressure p is assumed to satisfy the ideal gas equation of state

$$p = nk_B T \tag{2.9}$$

where $n = \sum n_i$. The temperature satisfies the equation

$$c_p \rho \frac{DT}{Dt} - \frac{Dp}{Dt} = -\nabla \cdot \vec{q} + \underline{\underline{\tau}} : \nabla \vec{v} + \sum_{i=1}^{N} n_k(\vec{v}_k - \vec{v}) \cdot \vec{F}_k - \sum_{i=1}^{N} \widetilde{H}_i \mathcal{N}_i . \tag{2.10}$$

Here c_p is the specific heat at constant pressure of the gas and $\widetilde{H}_i$ is the partial molar enthalpy in units $J/molecule$.

The spatial boundary conditions for Equations (2.2), (2.3), (2.5) and (2.10) are of Dirichlet, Neumann or mixed type. Table 2.1 gives typical no slip boundary conditions valid in the hydrodynamic regime for a flow with nonreacting surfaces. A description of a more general case can be found in [50].

variable	boundary		
	wall	inlet	outlet
n_i	$\vec{n} \cdot \vec{j}_i = 0$	$n_i = n_i(inlet)$ $\vec{n} \cdot \vec{j}_i = 0$	$\vec{n} \cdot \vec{j}_i = 0$
$\vec{v}$	$\vec{v} = 0$	$\vec{n} \cdot \vec{v} = v(inlet)$ $\vec{t} \cdot \vec{v} = 0$	$\vec{n} \cdot \nabla(\rho\vec{v}) = 0$ $\vec{t} \cdot \vec{v} = 0$
T	$T = T(wall)$	$T = T(inlet)$	$\vec{n} \cdot \vec{\varepsilon} = 0$

Table 2.1: Typical no slip boundary conditions. $\vec{n}$ and $\vec{t}$ are unit vectors normal and tangential to the boundary.

2.2.2 Generalized Forces

The balance equations (2.3), (2.4) and (2.5) may now be combined to derive a balance equation for the specific entropy of the flow. The advantage is that the change in entropy may be viewed as the results of a reversible (conservative) process and an irreversible process. To ensure that the physical condition of increasing entropy is satisfied, the irreversible entropy is expressed as a symmetric positive definite quadratic form acting on generalized forces. The coefficients of the quadratic form are the transport coefficients. This formulation is achieved by expressing the flux quantities as linear combinations of the generalized forces.

The change of the specific entropy $\hat{s}$ under transport is given by the balance equation

$$\rho \frac{D(\hat{s})}{Dt} = \frac{\rho}{T} \frac{D(\hat{u})}{Dt} + \frac{p\rho}{T} \frac{DV}{Dt} - \frac{\rho}{T} \sum_{i=1}^{N} \frac{\mu_i}{m_i} \frac{D\omega_i}{Dt} \tag{2.11}$$

where $\mu_i = \mu_i(T, n_1, ..., n_N) = k_B T \ln(n_i/n)$ is the chemical potential of the species i in units $J/molecule$ (under the assumption of an ideal gas mixture). Substitution of Equations (2.3), (2.4) and (2.5) in to Equation (2.11) leads to

$$\rho \frac{D(\hat{s})}{Dt} = -\nabla \cdot \left(\frac{1}{T}\vec{q} - \frac{1}{T} \sum_{i=1}^{N} \frac{\mu_i}{m_i}\vec{j}_i \right) - \frac{1}{T^2}\vec{q} \cdot \nabla T \tag{2.12}$$

$$+ \frac{1}{T}\underline{\underline{\tau}} : \nabla\vec{v} - \sum_{i=1}^{N} \frac{1}{m_i}\vec{j}_i \cdot \left(\nabla \left(\frac{\mu_i}{T} \right) - \frac{1}{T}\vec{F}_i \right) - \frac{1}{T} \sum_{i=1}^{N} \mu_i \mathcal{N}_i.$$

It is helpful to express the total energy flux $\vec{q}$ as the sum of the thermal energy flux $\vec{\varepsilon}$ (flux by thermal gradient and radiation) and the flux of energy carried with the diffusing molecules

$$\vec{q} = \vec{\varepsilon} + \sum_{i=1,N} n_i \widetilde{H}_i (\vec{v}_i - \vec{v}) = \vec{\varepsilon} + \sum_{i=1,N} \widetilde{H}_i \vec{j}_i / m_i \ . \tag{2.13}$$

Further, it is also convenient to introduce the partial molar entropy $\widetilde{S}_i$ which may be expressed as

$$\widetilde{S}_i = \frac{\widetilde{H}_i - \mu_i}{T} \ . \tag{2.14}$$

This relationship follows from the thermodynamical equation for the total enthalpy density h

$$h = \sum_{i=1}^{N} n_i \widetilde{H}_i = e + p = T \sum_{i=1}^{N} n_i \widetilde{S}_i + \sum_{i=1}^{N} \mu_i n_i \ . \tag{2.15}$$

The balance equation for specific entropy may then be expressed as

$$\rho \frac{D\hat{s}}{Dt} = \dot{S}^{rev} + \dot{S}^{irrev} \tag{2.16}$$

where

$$\dot{S}^{rev} = -\nabla \cdot \left(\frac{1}{T} \vec{q} - \frac{1}{T} \sum_{i=1}^{N} \frac{\mu_i}{m_i} \vec{j}_i \right) \tag{2.17}$$

and

$$\dot{S}^{irrev} = \frac{1}{T} \left(-\vec{\varepsilon} \cdot \vec{X}_1 + \underline{\underline{\tau}} : \underline{\underline{X}}_2 - \sum_{i=1}^{N} \vec{j}_i \cdot \vec{X}_3^i - \sum_{i=1}^{N} \mathcal{N}_i X_4^i \right) \ . \tag{2.18}$$

The X_k are the generalized forces defined as

$$\begin{aligned} \vec{X}_1 &= \frac{1}{T} \nabla T \\ \underline{\underline{X}}_2 &= \nabla \vec{v} \\ \vec{X}_3^i &= \frac{1}{m_i} \left(\nabla \mu_i + \widetilde{S}_i \nabla T - \vec{F}_i \right) \\ X_4^i &= \mu_i \ . \end{aligned} \tag{2.19}$$

The Onsager theory of irreversible thermodynamics [25] states that if the system is not too far from equilibrium, the fluxes J_l ($\vec{\varepsilon}$, $\underline{\underline{\tau}}$, $\vec{j}_i$ and $\mathcal{N}_i$) can be expressed as appropriate linear combinations of the generalized forces

$$J_l = \sum_{l'=1}^{2N+2} \alpha_{ll'} X_{l'} \ . \tag{2.20}$$

The $\alpha_{ll'}$ are the transport coefficients. This is somewhat loose use of notation as the $\alpha_{ll'}$ are perhaps better viewed as symmetric linear operators on the generalized forces. The irreversible entropy production may then be expressed as

$$\dot{S}^{irrev} = \frac{1}{T} \sum_{l=1}^{2N+2} \sum_{l'=1}^{2N+2} X_{l'} \alpha_{ll'} X_l \qquad (2.21)$$

Onsager's theorem gives the structure of the fluxes in Equation (2.20) in detail (see [25], [34]) which ensures that $\dot{S}^{irrev}$ is non-negative. The structure of the momentum current is the first result of the theory. For a Newtonian fluid (i.e., a fluid for which stress is linear in force), the viscous stress tensor is

$$\underline{\tau} = \rho\nu \left(\nabla \vec{v} + (\nabla \vec{v})^\dagger \right) + \left(\lambda - \frac{2}{3}\rho\nu \right) (\nabla \cdot \vec{v}) \, \underline{I}. \qquad (2.22)$$

Here ν is the kinematic viscosity (in units m^2/s), and λ is the bulk viscosity. For this choice of $\underline{\tau}$ the fluid equations are known as the Navier-Stokes equations.

The heat flux $\vec{\varepsilon}$ and the diffusive flux $\vec{j}_i$ are most conveniently expressed in terms of the variable $\vec{d}_i$, defined as

$$\vec{d}_i \; := \; \frac{1}{p} \left(\rho_i \vec{X}_3^i - \frac{\rho_i}{\rho} \sum_{j=1}^N \rho_j \vec{X}_3^j \right) \qquad (2.23)$$

$$= \; \nabla \left(\frac{n_i}{n} \right) + \left(\frac{n_i}{n} - \frac{\rho_i}{\rho} \right) \nabla \ln p - \frac{1}{p}\frac{\rho_i}{\rho} \left(\frac{\rho}{m_i} \vec{F}_i - \sum_{j=1}^N n_j \vec{F}_j \right).$$

The second equation follows from the fact that the $\vec{X}_3^i$, under the assumptions of an ideal gas mixture, may be expressed as[1]

$$\vec{X}_3^i = \frac{\nabla p_i}{\rho_i} - \frac{\vec{F}_i}{m_i}. \qquad (2.24)$$

---

[1] We first evaluate $\vec{X}_3^i$ with the help of

$$\nabla \mu_i(p, T, n_1, ..., n_N) = -\overline{S}_i \nabla T + \frac{1}{n}\nabla p + kT \nabla \ln \left(\frac{n_i}{n} \right)$$

where we used $\mu_i = kT \ln(n_i/n)$ and

$$\left(\frac{\partial \mu_i}{\partial T} \right)_{p,n_j} = -\widetilde{S}_i \quad , \quad \left(\frac{\partial \mu_i}{\partial p} \right)_{T,n_j} = \frac{1}{n} \quad , \quad \left(\frac{\partial \mu_i}{\partial n_j} \right)_{T,p} = \frac{kT}{n_i}\delta_{ij} - \frac{kT}{n}$$

which corresponds to the assumption of an ideal gas mixture. We then have

$$\vec{X}_3^i = \frac{1}{m_i} \left(\frac{1}{n}\nabla p + kT \nabla \ln \left(\frac{n_i}{n} \right) - \vec{F}_i \right) = \frac{\nabla p_i}{\rho_i} - \frac{F_i}{m_i}.$$

This transformation makes the linear dependence $\sum_{i=1,N} \vec{d_i} = 0$ explicit, just as with the corresponding diffusion current $\vec{j_i}$. Also, the concentration gradient, the pressure gradient and the external force are separated as components of the generalized force.

The Onsager theory then gives the heat current as a linear combination of the vector forces

$$\vec{\varepsilon} = -\kappa \nabla T - p \sum_{j=1}^{N} \frac{D_i^T}{\rho_i} \vec{d_i} \qquad (2.25)$$

D_i^T is the multicomponent thermal diffusion coefficient in units $Kg/(ms)$ and κ is the multicomponent thermal conductivity in units $Kg/(Km)$ up to a correction which accounts for the concentration gradients due to the thermal diffusion in a thermal conductivity measurement (see [97] for a discussion). The predictions of D_i^T from kinetic theory depend strongly on the assumed intermolecular potential (see [50]).

The diffusive current is given by a different linear combination of the vector forces

$$\vec{j_i} = -\frac{D_i^T}{T} \nabla T + \frac{n^2}{\rho} \sum_{j=1}^{N} m_i m_j D_{ij} \vec{d_j} \qquad (2.26)$$

D_{ij} are the multicomponent diffusion coefficients in units m^2/s, and D_i^T are the same multicomponent thermal diffusion coefficients as in the heat current in units $Kg/(ms)$. The D_{ij} are concentration dependent quantities, but they can in principle be expressed through the binary diffusion coefficient $\mathcal{D}_{ij}$ with the help of the Stefan-Maxwell relations [50], [34]. For a binary mixture, D_{ij} is identical with $\mathcal{D}_{ij}$, and the diffusion current (neglecting external forces) is

$$\vec{j_1} = -\frac{D_1^T}{T} \nabla T - \frac{n^2}{\rho} m_1 m_2 \mathcal{D}_{12} \left(\nabla \left(\frac{n_1}{n} \right) + \left(\frac{n_1}{n} - \frac{\rho_1}{\rho} \right) \nabla \ln p \right) \qquad (2.27)$$

with $\vec{j_2} = -\vec{j_1}$, $\mathcal{D}_{12} = \mathcal{D}_{21}$, $D_1^T = -D_2^T$. Often, the pressure gradient is neglected and the current can conveniently be expressed as a gradient of the mass fraction

$$\vec{j_1} = -\frac{D_1^T}{T} \nabla T - \rho \mathcal{D}_{12} \nabla \omega_1 . \qquad (2.28)$$

Table 2.2 gives an overview of the transport phenomena and their driving forces in the hydrodynamic approximation.

2.3   Near Hydrodynamic Flows

The near hydrodynamic regime consist of flows which may still be modeled by continuum equations, but for which kinetic effects begin to appear.

current	generalized forces					
	tensor	vector			scalar	
	$\nabla \vec{v}$	$\nabla \ln T$	$\nabla \ln p$	$\nabla (n_i/n)$	$\nabla \cdot \vec{v}$	μ_i
$\underline{\underline{\tau}}$	shear stress				bulk stress	
$\vec{\varepsilon}$		thermal conduction	unnamed	Dufour effect		
$\vec{j_i}$		thermal diffusion	pressure diffusion	ordinary diffusion		
$\mathcal{N}_i$						chemical reaction

Table 2.2: Transport phenomena in bulk flow related to the different forces

One of the first successes of kinetic theory was to provide good estimates for viscosity, thermal conductivity and diffusivity for these flows based on molecular interaction. These results provided insight into the physical nature of hydrodynamic flows, as well as useful models. The most basic of these are outlined below.

Further insight is obtained from the Chapman-Engskog derivation of the fluid equations from the Boltzmann equation. By including the associated slip boundary conditions, instead of the more standard no-slip conditions given in Table 2.1, the validity of the equations may be extending into the near hydrodynamic regime. The appropriate boundary conditions are presented below.

Finally, higher order terms in the Chapman-Engskog expansion may be included to provide a more detailed description of the diffusive and heat currents and the viscous stress tensor. The resulting Burnett equations have terms with higher order derivatives without clear physical interpretation, and which require additional boundary conditions. For this reason they are rarely used in gas flow modeling. However, second order effects for the diffusive current may still be introduced without the problems associated with the Burnett equations through the use of an effective diffusion coefficient in Equation (2.26). This idea is explored in more detail in Section 4.1.2.

2.3.1 Estimates of Transport Coefficients

Kinetic theory provides detailed expressions for the viscosity, thermal conductivity and diffusion coefficients [34], [97], [50]. These depend on the molecular interaction model; more specifically they are defined through integrals of functions of the collision scattering angle $\chi(b)$ (see Equation (1.20)) with respect to the impact parameter b. Order of magnitude estimates can be given based on molecular mass m, temperature T, collision diameter d and the number density $n = p/(k_B T)$. The mean free path, mean thermal speed and specific heat per molecule are estimated as

$$\lambda_{est} \sim \frac{1}{d^2 n} \quad , \quad \bar{v}_{th} \sim \sqrt{\frac{k_B T}{m}} \quad , \quad c_p \sim k_B \ . \tag{2.29}$$

These are in turn used to estimate the transport coefficients. The thermal conductivity is estimated as

$$\kappa_{est} \sim c_p n \lambda \bar{v} = \frac{k_B}{d^2} \sqrt{\frac{k_B T}{m}} \qquad \left[\frac{W}{mK} \right] \ . \tag{2.30}$$

The dynamic viscosity $\mu = \rho \nu$ is estimated as

$$\mu_{est} \sim m n \lambda \bar{v} = \frac{m}{d^2} \sqrt{\frac{k_B T}{m}} \qquad \left[\frac{Kg}{ms} \right] \ , \tag{2.31}$$

while the self-diffusion (binary diffusion where both components have identical properties) is

$$\mathcal{D}_{est} \sim \lambda \bar{v} = \frac{k_B T}{p} \frac{1}{d^2} \sqrt{\frac{k_B T}{m}} \qquad \left[\frac{m^2}{s} \right] \ . \tag{2.32}$$

The hard sphere model predicts numerical coefficients which improve the accuracy of the above expressions. For a single species gas they are

$$\lambda_{HS} = \frac{1}{\pi \sqrt{2}} \lambda_{est} \quad , \quad \kappa_{HS} = \frac{f}{2} \frac{1}{\sqrt{\pi}} \frac{25}{32} \kappa_{est} \tag{2.33}$$

$$\mu_{HS} = \frac{1}{\sqrt{\pi}} \frac{5}{16} \mu_{est} \quad , \quad \mathcal{D}_{HS} = \frac{1}{\sqrt{\pi}} \frac{3}{8} \mathcal{D}_{est}$$

where f is the number of degrees of freedom of the gas (e.g., rotational, vibrational, etc.).

Important for numerical calculations are the relaxation times for the different flow variables. They determine the number of time steps required to reach a stationary state in a Monte Carlo calculation, as well as the relaxation parameters used in the numerical solution of the hydrodynamic differential equation. The relaxation time scale for the flow is

$$\tau_v = \frac{\rho \mathcal{L}^2}{\mu} \tag{2.34}$$

with $\mathcal{L}$ the typical length scale of the reactor. For the temperature it is

$$\tau_T = \frac{\rho \mathcal{L}^2 c_p}{\kappa} \ , \tag{2.35}$$

and for the diffusion

$$\tau_D = \frac{\mathcal{L}^2}{\mathcal{D}}. \tag{2.36}$$

As an example, these time scales can be estimated for the flow of an equimolar mixture of SiH_4 and H_2 in a UHV-CVD reactor at $p = 1\ Pa$ and $T = 600°K$ with $\mathcal{L} = 0.1\ m$, $\rho = 3.4 \times 10^{-6}\ Kg/m^3$, $\mu = 2.5 \times 10^{-5}\ Kg/(ms)$, $\kappa = 0.26\ W/(mK)$, $\mathcal{D} = 18\ m^2/s$, $c_p = 5 \times 10^3\ J/(KgK)$ and the collision diameters from Table 2.3 as

$$\tau_v = 1.4 \times 10^{-3}s \quad , \quad \tau_T = 0.65 \times 10^{-3}s \quad , \quad \tau_D = 0.55 \times 10^{-3}s. \tag{2.37}$$

2.3.2 Slip Boundary Conditions

The hydrodynamic equations contain terms for the mass, momentum and energy flux expressed in terms of gradients of mass fraction, velocity and temperature. A rigorous treatment of the Boltzmann equation using the Chapman-Engskog expansion (see Section 1.2.2) for a multiple species gas leads to similar results [21]. This method shows that the gradient terms are in fact just the first order terms of the expansion (the zeroth order terms have no heat flux and the stress tensor is simply the static pressure given by the ideal gas law). The expansion parameter is the Knudsen number $Kn = \lambda/\mathcal{L}$.

For a single species gas, boundary conditions have been derived rigorously for the Chapman-Engskog expansion from the Boltzmann equation boundary conditions [20], [22]. The resulting equations include first order corrections in Kn to the no slip boundary conditions of Table 2.1. For two species flows, thermodynamic arguments similar to the Onsager theory have been used [92], [10] to obtain slip boundary conditions. We propose here a natural extension of these results to a multiple species gas.

The resulting boundary condition for the stream velocity $\vec{v}$ at a wall with unit normal vector $\vec{n}$ is

$$\begin{aligned}
\vec{t} \cdot \vec{v}\big|_{z=0} &= \zeta^{vv}\vec{n} \cdot \nabla(\vec{t} \cdot \vec{v}) + \zeta^{vT}\vec{t} \cdot \frac{\nabla T}{T} + \sum_{i=1}^{N} \zeta_i^{vp}\vec{t} \cdot \vec{X}_3^i \\
&= \zeta^{vv}\vec{n} \cdot \underline{\underline{X}}_2 \cdot \vec{t} + \zeta^{vT}\vec{t} \cdot \vec{X}_1 + \sum_{i=1}^{N} \zeta_i^{vp}\vec{t} \cdot \vec{X}_3^i \\
\vec{n} \cdot \vec{v}\big|_{z=0} &= 0
\end{aligned} \tag{2.38}$$

where $\vec{t}$ is an arbitrary tangent vector. The first term is called the velocity slip effect. The next term is the thermal creep effect, and the last term is the diffusive creep effect. There are no other vector forces which can be constructed from $\vec{X}_1$, $\underline{\underline{X}}_2$, $\vec{X}_3^i$ from (2.19) and $\vec{n}$.

The temperature boundary condition can be formulated as

$$T_{wall} - T|_{z=0} = \zeta^{TT}\vec{n} \cdot \frac{\nabla T}{T} + \sum_{i=1}^{N} \zeta_i^{Tp}\vec{n} \cdot \vec{X}_3^i \qquad (2.39)$$

$$= \zeta^{TT}\vec{n} \cdot \vec{X}_1 + \sum_{i=1}^{N} \zeta_i^{Tp}\vec{n} \cdot \vec{X}_3^i$$

The first term is called the temperature jump effect. The other term represents the effects of the normal concentration gradients or external forces in normal direction. There are two other scalar forces which can be constructed from $\underline{X}_2$ and $\vec{n}$ which were neglected here. These are $\nabla \cdot \vec{v} = \underline{\underline{I}} : \underline{\underline{X}}_2$ and $\vec{n} \cdot \nabla(\overline{\vec{v} \cdot \vec{n}}) = \vec{n} \cdot \underline{\underline{X}}_2 \cdot \vec{n}$

From the Chapman-Engskog expansion it follows that the slip coefficients ζ should be proportional to the mean free path λ. A summary of various analytic techniques which have been used to evaluate the slip coefficients can be found in Cercignani's book [20]. For example, for a wall with diffuse reflection, the BGK collision model leads to $\zeta^{vv} = 1.1466\lambda$. In Chapter 4, a detailed computational study of the velocity and thermal slip, as well as the thermal and diffusive creep coefficients, is presented. Here the Direct Simulation Monte Carlo method, described in the next chapter, is used to determine the coefficients and their dependence on species and partial pressure for mixtures.

2.4   Transition Regime Flows

We now turn to transition regime flows. These may be roughly defined as gas flows for which the Knudsen number lies in the range $0.1 < Kn < 1$. Here the particle velocity distribution function may be so far from the equilibrium Maxwellian distribution that it may no longer be viewed as a perturbation of the equilibrium. The validity of the Chapman-Enskog expansion and the resulting Navier-Stokes fluid equations becomes questionable. Similarly, the Onsager theory also breaks down because the transition regime flows may be considerably far from thermodynamic equilibrium. The deviations of the Navier-Stokes solution from the true Boltzmann equation solution may be quite significant for these flows; estimates of quantities such as heat flux may be off by orders of magnitude should the continuum fluid dynamics equations be used.

For transition regime flows the full Boltzmann equation (see Section 1.2) must be solved. The next chapter is devoted to a Monte Carlo method which has been developed to compute these flow. Before describing the details of the numerics, however, we will address two aspects of the modeling of a rarefied gas flow. The first concerns the description of the molecular interaction which enters the Boltzmann equation through the factor $B(V, \chi)$ in Equation (1.16). The most physically realistic interaction potentials have the disadvantage of being somewhat computationally awkward; thus we consider a simplified interaction which nonetheless captures most of the relevant physics. The second issue addressed here concerns the choice of boundary conditions, specifically for inlets and outlets, which arise in the simulation of reactor flows.

2.4.1 Scattering Angle Models

The physics of two point molecules acting under the influence of a spherical potential $\Phi(r)$ was described in Section 1.2. Based on the conservation of linear and angular momentum, a relationship was derived which gives the angle $\theta(b)$ between the pre-collision relative velocity vector and the normal vector about which the particles scatter as a function of the impact parameter b. The scattering angle (the angle between the pre-collision and post-collision relative velocities) is defined as $\chi(b) = \pi - 2\theta(b)$. From Equation (1.20), $\chi(b)$ can be seen to satisfy

$$\chi = \pi - 2b \int_{r_{min}}^{\infty} \frac{dr}{r^2 \sqrt{1 - 2\Phi(r)/m_r V_r^2 - b^2/r^2}} \qquad (2.40)$$

where r_{min} is the distance of closest approach given by solving

$$b^2 = r_{min}^2 \left(1 - \frac{2\Phi(r_{min})}{m_r V_r^2} \right) . \qquad (2.41)$$

Here m_r is the reduced mass of the two colliding particles as defined in Equation (1.17) and V_r is the magnitude of the relative velocity of the colliding pair.

In a Monte Carlo particle simulation, when two particles collide it is necessary to determine an appropriate scattering angle. This may be done by randomly choosing a value for the impact parameter b uniformly from the range $0 \leq b \leq b_{max}$, where b_{max} is the cut-off value for the potential (i.e., particles separated by a distance greater than b_{max} have no interaction). The fact that b is uniformly distributed means that prior to the collision, the particles have no influence on each other - thus they are equally likely to collide head on ($b = 0$) as to have a minimally glancing collision ($b = b_{max}$). Once a value for b is selected, the corresponding scattering angle $\chi(b)$ may be computed by evaluating the integral in Equation (2.40).

For realistic potentials $\Phi(r)$ the integral in Equation (2.40) does not have a closed form expression. Therefore it is necessary to compute it numerically. This is rather expensive given the great number of collision which occur in the course of a simulation. An alternative is to tabulate the integral; however, this is also somewhat impractical because of the dependence of the integral on both b and V_r (which is unbounded). Moreover, such a table would be required for every pair of species, which would add to the expense for a multiple species gas.

The VHS Model

Bird [13] proposed an effective means of avoiding this computational problem. He observed that the simulation results depended strongly on the total collision cross sections (as functions of V_r), which are integrals of the scattering angle $\chi(b)$, but only weakly on the actually form of the scattering angle itself. Bird then proposed his Variable Hard Sphere (VHS) model which quite accurately captures the velocity dependence of the viscosity cross section, while using the easily computable scattering of hard sphere collisions. This is achieved by making the diameter of the colliding particles an appropriate function of the relative velocity V_r.

To better understand this model and to motivate the introduction of an improved model, M1, described below, we present now some results from kinetic theory. It is well known that the transport properties of a gas can be expressed by integrals over the molecular cross section $b(\chi)$ [34], [97]. The following expressions for the properties for pure species have been derived from the Chapman-Engskog expansion. For the viscosity

$$\mu = \frac{5}{16} \frac{m}{\sqrt{\pi}} \sqrt{\frac{k_B T}{m}} \frac{1}{d^2 \Omega^{(2,2)*}} \tag{2.42}$$

for the heat conductivity

$$\kappa = \frac{f}{2} \frac{25}{32} \frac{k_B}{\sqrt{\pi}} \sqrt{\frac{k_B T}{m}} \frac{1}{d^2 \Omega^{(2,2)*}} \tag{2.43}$$

and for the diffusion coefficient

$$D = \frac{3}{8} \frac{1}{\sqrt{\pi}} \frac{k_B T}{p} \sqrt{\frac{k_B T}{m}} \frac{1}{d^2 \Omega^{(1,1)*}} \ . \tag{2.44}$$

Here the same notation as in Section 2.3.1 is used. Usually not all of the degrees of freedom of the molecules in a gas are excited at the same time. For this reason $f = 2c_v/k_B$ may be substituted into Equation (2.43), where $c_v(T)$ is the specific heat at constant volume. For the specific heat, polynomial approximations exist [50]. The Ω-collision integrals [34] are moments over distributions of relative velocities and total cross section $Q^{(l)}$ at temperature T:

$$\Omega^{(l,s)}(T) = \sqrt{\frac{k_B T}{2\pi m_r}} \int_0^\infty e^{-\gamma^2} \gamma^{2s+3} Q^{(l)}(V_r) d\gamma \tag{2.45}$$

$$\gamma^2 = \frac{m_r V_r^2}{2k_B T} \tag{2.46}$$

$$Q^{(l)} = \int_0^\infty \left(1 - \cos{}^l \chi(b)\right) b \, db \tag{2.47}$$

The Ω-integrals are thus defined by specifying $\chi(b)$. $Q^{(1)}(V_r)$ is called the diffusion cross section, while $Q^{(2)}(V_r)$ is known as the viscosity cross section.

The reduced integrals Ω^* which appear in Equations (2.42), (2.43), and (2.44) are the Ω-integrals normalized by the corresponding integrals for the hard sphere model:

$$\Omega^{(l,s)*} = \frac{\Omega^{(l,s)}}{\Omega_{HS}^{(l,s)}} \qquad \Omega_{HS}^{(l,s)} = \sqrt{\frac{k_B T}{2\pi m_r}} \frac{(s+1)!}{2} \left(1 - \frac{1}{2}\frac{1+(-1)^l}{1+l}\right) \pi d_{HS}^2 \cdot \tag{2.48}$$

Here d_{HS} is the diameter of the hard sphere molecule. For Ω^* polynomial approximations can be found in [50].

We now describe the details of the Variable Hard Sphere model. The scattering angle $\chi(b)$ is taken to be the same as well known hard sphere scattering angle, which is based on a potential $\Phi_{HS}(r)$ which is infinite for $r < d_{HS}$ and zero for $r > d_{HS}$. The VHS model simply replaces the hard sphere diameter d_{HS} with a variable diameter $d_{VHS}(V_r)$ which is function of the relative velocity. The resulting VHS scattering angle $\chi(b)$ as a function of impact parameter b is

$$\chi(b) = 2 \arccos\left(\frac{b}{d_{VHS}(V_r)}\right) \qquad b < d_{VHS}(V_r) \tag{2.49}$$

The probability distribution of the impact parameter is then

$$b(\chi) = d_{VHS} \cos(\chi/2), \tag{2.50}$$

which shows that the scattering angle has indeed an uniform distribution over the sphere. This is the main simplification of the VHS-model. The corresponding diffusion and viscosity cross sections from Equation (2.47) are

$$Q_{VHS}^{(1)}(V_r) = \pi d_{VHS}^2(V_r) \quad , \qquad Q_{VHS}^{(2)}(V_r) = \frac{2}{3}\pi d_{VHS}^2(V_r) . \tag{2.51}$$

The question remains as to how to choose $d_{VHS}(V_r)$. If it is set constant at the nominal hard sphere diameter d_{HS}, then this is simply the hard sphere model. This model, however, does not adequately model the transport coefficients of a real gas. A careful choice of $d_{VHS}(V_r)$ will result in a much more accurate representation of the transport parameters. Bird [13] proposes to set

$$d_{VHS}(V_r) \;=\; d_{VHS}^{\text{ref}} \sqrt{\left(\frac{2k_B T_{\text{ref}}}{m_r V_r^2}\right)^{\omega-1/2} \frac{1}{\Gamma(5/2-\omega)}}$$

$$d_{VHS}^{\text{ref}} \;=\; \sqrt{\frac{15\sqrt{mk_B T_{\text{ref}}/\pi}}{2(5-2\omega)(7-2\omega)\mu(T_{\text{ref}})}} \,. \tag{2.52}$$

Here $\mu(T_{ref})$, T_{ref} and ω are parameters.

With this choice for $d_{VHS}(V_r)$ and the scattering angle of Equation (2.49), the reduced $\Omega^{(2,2)*}$-integral can now be evaluated. The resulting VHS-viscosity is

$$\mu_{VHS}(T) = \mu(T_{ref}) \left(\frac{T}{T_{ref}}\right)^{\omega} \,. \tag{2.53}$$

The meaning of the parameters in Equation (2.52) is now clear. T_{ref} is a reference temperature, $\mu(T_{ref})$ is the viscosity of the gas at that temperature, and ω is an exponent which describes the temperature dependence of the viscosity. For inverse power law potentials ($\Phi(r) = cr^{-\eta}$, $\eta \geq 4$), it can be shown [21] the viscosity defined by Equation (2.42) has exactly the form of Equation (2.53). This is therefore a reasonable model for the viscosity. The heat conductivity and the self-diffusion coefficient have a related temperature dependence given by

$$\kappa_{VHS}(T) \;=\; \frac{5}{2}\frac{c_v(T)}{m}\,\mu_{VHS}(T) \tag{2.54}$$

$$D_{VHS}(T) \;=\; \frac{7-\omega}{5}\frac{k_B T}{m\,p}\,\mu_{VHS}(T) \tag{2.55}$$

The VHS model is completed by specifying the parameters $\mu(T_{ref})$ and ω. They can be obtained from a fit of $\mu_{VHS}(T)$ to experimental data of the temperature dependence of the single species viscosity. Since such data are hardly available for all molecular species, an alternative procedure is to fit $\mu_{VHS}(T)$ to the Lennard-Jones model description of $\mu(T)$, which is known to represent the experimental values very well. The Lennard-Jones parametrization consists of the two parameters σ and ϵ/k_B and the associated molecular potential $\Phi(r) = 4\epsilon\left((\sigma/r)^{12} - (\sigma/r)^6\right)$. Tabulated values for the collision integrals can be found in [34] and [50]. Table 2.3 contains the Lennard-Jones parameters, the fitted VHS parameters and values for the viscosity and self-diffusion coefficients for a variety of molecular species.

According to the above description of transport properties, the ratio of thermal conductivity and viscosity (at fixed temperature) is always a constant, $\kappa/\mu = 5/2\,c_v(T)/m$. This confirmed by experiment [97] and fulfilled by the models of the molecular potential to the degree that it can be

	M $[\frac{g}{mole}]$	σ $[\mathring{A}]$	ϵ/k_B $[^\circ K]$	μ $[\frac{10^{-5}Kg}{ms}]$	ω	D_{LJ} $[\frac{cm^2}{s}]$	D_{VHS} $[\frac{cm^2}{s}]$	λ_{VHS} $[\mu m]$
Ar	39.94	3.330	136.5	2.15	0.74	0.160	0.135	0.054
AsH₃	77.95	4.145	259.8	1.46	0.85	0.056	0.045	0.023
CN	26.08	3.856	75.0	1.52	0.69	0.175	0.149	0.049
CH₄	16.04	3.746	141.4	1.06	0.75	0.197	0.165	0.041
TMG	114.83	5.680	398.0	0.63	0.95	0.024	0.019	0.010
H₂	2.016	2.920	38.0	0.83	0.68	1.233	1.055	0.096
HF	20.01	3.138	330.0	1.14	0.91	0.172	0.135	0.034
HCl	36.46	3.339	344.7	1.33	0.94	0.133	0.110	0.028
H₂O	18.01	2.605	572.0	1.20	0.99	0.197	0.151	0.034
N	14.01	3.298	71.4	1.55	0.68	0.329	0.283	0.068
N₂	28.02	3.621	97.5	1.65	0.68	0.176	0.151	0.051
NH₃	17.03	2.920	481.0	1.01	0.97	0.177	0.136	0.030
O	16.00	2.750	80.0	2.32	0.69	0.432	0.369	0.094
O₂	32.00	3.458	107.4	1.91	0.72	0.178	0.151	0.054
S	32.06	3.839	847.0	0.62	0.98	0.056	0.045	0.014
Si	28.09	2.910	3036.0	0.77	0.70	0.073	0.069	0.023
SiF₄	104.09	4.880	171.9	1.47	0.78	0.042	0.035	0.022
SiH₂	30.10	3.803	133.1	1.44	0.74	0.143	0.120	0.041
SiH₄	32.12	4.084	207.6	1.07	0.78	0.100	0.081	0.027
Si₂H₄	60.21	4.601	312.6	0.95	0.90	0.047	0.037	0.016
Si₂H₆	62.23	4.828	301.3	0.89	0.90	0.043	0.034	0.015
Si₃H₈	92.33	5.562	331.2	0.78	0.92	0.025	0.020	0.011
WF₆	297.84	5.210	338.0	1.58	0.93	0.016	0.012	0.012
TEOS	208.33	3.542	93.3	4.85	0.70	0.069	0.059	0.054

Table 2.3: This table contains the molecular mass and the Lennard-Jones parameter of various molecular and atomic species. These parameter are taken from the CHEMKIN data base [41]. The viscosity at a temperature of $T = 273^\circ K$ is calculated in the Lennard-Jones model with the collision integral interpolation formula from [50]. The VHS-temperature exponent is obtained from a fit to the viscosity in the interval $[250^\circ K, 1000^\circ K]$. The self-diffusion coefficients are calculated in the Lennard-Jones and VHS-model, respectively. Finally, the VHS mean free path is given at normal pressure of 10^5 Pa. TMG is the species $Ga(CH_3)_3$ and TEOS is $Si(OC_2H_5)_4$.

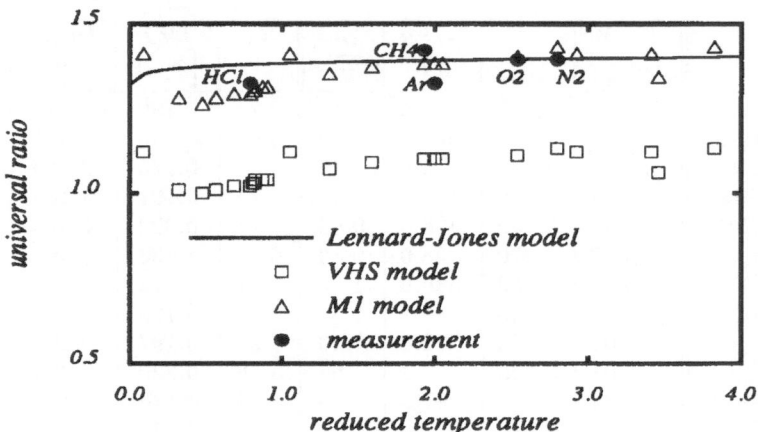

Figure 2.2: Universal ratio $\rho D/\mu$ plotted against the reduced temperature $T^* = T_{ref}/(\epsilon/k_B)$ at $T_{ref} = 273°\,K$. Shown are values from the Lennard-Jones model, from the VHS-model and from the M1 model. Also shown are experimental results from [97].

approximated as spherically symmetric. Another combination, namely

$$\frac{D\rho}{\mu} = \frac{6}{5} \frac{\Omega^{(2,2)*}(T)}{\Omega^{(1,1)*}(T)} \tag{2.56}$$

is also universal. The Lennard-Jones collision integrals yield an almost constant value around 1.32 whereas the VHS-model gives the constant $(7-\omega)/5$. Figure 2.2 show the values of this universal ratio for the Lennard-Jones model, VHS model and for some experimental data. The Lennard-Jones values are very close to the experimental data, whereas the VHS-values show a systematic deviation. Since the viscosity is captured very well by the VHS model, the deviation comes entirely from the VHS diffusion which is systematically about 30% too small. The reason lies in the grossly simplified differential cross section of the sphere, which implies more backwards oriented scattering than a more realistic description which impedes the diffusion. This observation motivates the introduction of a modification to the VHS model which we now discuss. The goal is to maintain the correct viscosity behavior while more adequately modeling the diffusion.

The M1 Model

The VHS model is fitted to provide very good values for the viscosity. For the diffusion, however, it provides values with a systematic error because the potential model of the variable sphere is too simple to describe the other transport properties at the same time. It is of course possible to modify the VHS-model in such a way that the diffusion coefficients are fitted for the price that the viscosity has a systematic error. This was done by Nanbu [71]. In order to obtain a model which fits both transport parameters well, we must return to the description of the differential cross section.

The goal is to find an expression for $\chi(b)$ similar to Equation (2.49) which does not require the numerical evaluation of an integral, but which more adequately captures the true scattering behavior of molecules. It is known that for the repulsive part of the molecular potential, the dependence of the scattering angle on the impact parameter is almost linear [34]. Based on this, we propose the following ansatz for the scattering angle:

$$\chi(b) = \pi \left(1 - \frac{b}{d_{M1}(V_r)} \right) . \tag{2.57}$$

We call this the M1 model. The diameter d_{M1} should have the same velocity dependence as d_{VHS}, but a larger reference diameter. It is defined as

$$d_{ref,M1} = \sqrt{\frac{4}{3}} d_{ref,VHS} \tag{2.58}$$

The total cross sections for this model from Equation (2.47) are

$$Q_{M1}^{(1)}(V_r) = \left(1 - \frac{4}{\pi^2} \right) \pi d_{M1}^2(V_r) , \quad Q_{M1}^{(2)}(V_r) = \frac{1}{2} \pi d_{M1}^2(V_r) . \tag{2.59}$$

The constant factor in Equation (2.58) has been chosen so that the viscosity cross sections $Q^{(2)}$ are equal for both models. This means that the M1 model will have the same viscosity behavior as the VHS model. The diffusion cross section of the M1 model is however smaller than that of the VHS model

$$Q_{M1}^{(1)} = \frac{4}{3} \left(1 - \frac{4}{\pi^2} \right) Q_{VHS}^{(1)} \approx 0.8 \, Q_{VHS}^{(1)} , \quad Q_{M1}^{(2)}(V_r) = Q_{VHS}^{(2)}(V_r) . \tag{2.60}$$

Hence the M1 model can be fitted to the viscosity data while simultaneously providing improved diffusion coefficients. Figure 2.2 also shows the values for the ratio $D\rho/\mu$ for the M1 model. The M1 model tends to overestimate the diffusion coefficient, but the values are much closer to the Lennard-Jones values and the experimental data than the VHS model.

The advantage of the M1 model over the VHS model can also be seen by directly comparing the scattering angles with the results for the Lennard-Jones model. Figure 2.3 (a) contains the scattering angle from a Lennard-Jones potential with parameters of argon $\sigma = 3.542 Å$ and $\epsilon/k_B = 93.3° K$ calculated from Equation (2.40) for different energies. It also shows the linear scattering angle from the M1 model with a temperature exponent

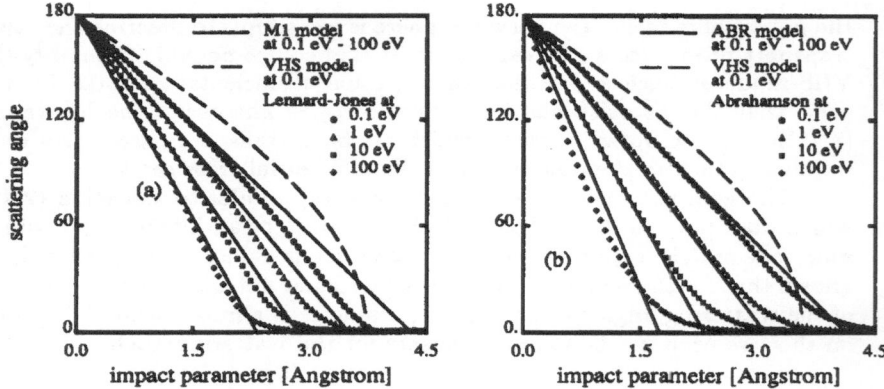

Figure 2.3: The scattering angle vs impact parameter obtained using different potential models.

$\omega = 0.6785$ for the same energies. This value of ω was choosen to match the energy dependence of the Lennard-Jones scattering angle as closely as possible. Finally it shows the scattering angle for the VHS model with the same temperature exponent at the lowest of the energies. The good agreement between the Lennard-Jones and M1 scattering angles can be observed.

The Lennard-Jones potential accurately describes the interaction of particles with kinetic energies corresponding to standard operating temperatures in micro-electronic processing equipment. There are some processes, however, for which certain particles may have kinetic energies well above the usual thermal energies. For example, in a sputtering reactor, the particles which are ejected from the source leave with an energy much higher than that of the background gas particles. In this case, the distance of closest approach during a collision between a sputtered molecule and a background molecule is much smaller than for a collision between thermally activated molecules. For these high energy collisions, the Lennard-Jones potential is known to be too "hard" and does not provide a physically realistic description. Thus it is necessary to find another model which more adequately describes the inner part of the potential probed by the high energy collisions. A suitable choice is the Abrahamson potential

$$\Phi(r) = A \exp\left(-\beta r\right) . \tag{2.61}$$

Values for argon are $A = 5942eV$ and $\beta = 3.663 \overset{\circ}{A}^{-1}$ [69]. Figure 2.3 (b) contains the scattering angle for the Abrahamson model as calculated from Equation (2.40). The potential is "softer" than the Lennard-Jones model, and as a consequence, it has a stronger energy dependence. This is reflected in the VHS and M1 models through the use of a larger temperature exponent, in this case $\omega = 0.75$. Again, the M1 scattering angle more accurately fits the Abrahamson model scattering angle than does the VHS scattering angle.

2.4.2 Boundary Conditions

Wall Interaction

As discussed in Chapter 1, gas-surface interaction is one of the least understood phenomena in chemistry. For non-absorbing, non-reacting surfaces, a simple accommodation coefficient model is usually used. Although not physically realistic, this model has proven to be as accurate in reproducing experimental results as more complicated models, and the simplicity makes it computationally attractive.

The idea is that a certain percentage α (the accommodation coefficient) of the particles which collide with a surface will thermally accommodate fully with the wall (diffuse reflection), while the remaining particles will gain no thermal information from the collision (specular reflection). Let T_w be the temperature of the surface, $\vec{n}$ be the outward unit normal to the surface, and $\vec{t}_1$ and $\vec{t}_2$ be two orthonormal tangent vectors. Furthermore let $\vec{u}_w$ be the velocity of the surface. For many applications the boundary walls are stationary, so $\vec{u}_w = 0$; occasionally a surface may have a tangential motion, such as a rotating cylinder. For steady problems, a non-zero normal velocity is rare. If $\vec{v}$ is the pre-collision (incoming) particle velocity and $\vec{v}'$ is the post-collision velocity, then specular reflection is calculated as

$$\begin{aligned} \vec{v}' \cdot \vec{n} &= (\vec{u}_w - \vec{v}) \cdot \vec{n} \\ \vec{v}' \cdot \vec{t}_i &= (\vec{u}_w - \vec{v}) \cdot \vec{t} \qquad i = 1, 2 \, . \end{aligned} \tag{2.62}$$

Diffuse reflection assumes that a particle striking the surface loses all memory of its incoming velocity and accommodates completely to the wall temperature and velocity. The particle therefore scatters from the surface according to a flux-weighted Maxwellian equilibrium distribution with temperature T_w and mean velocity $\vec{u}_w$. This means that in the normal direction to the wall the velocity v'_n is sampled from the distribution

$$f(v'_n) = c \, v'_n \exp\left(-\frac{(v'_n - \vec{u}_w \cdot \vec{n})}{2T_w}\right) \qquad v'_n > 0 \tag{2.63}$$

where c is the normalization constant

$$c = \left[\int_0^\infty v'_n \exp\left(-\frac{(v'_n - \vec{u}_w \cdot \vec{n})}{2T_w}\right) \, dv'_n\right]^{-1} .$$

The two tangential velocities are sampled from standard Maxwellians with temperature T_w and mean velocity $\vec{u}_w \cdot \vec{t}_1$ or $\vec{u}_w \cdot \vec{t}_2$.

Inflow and Outflow

Most flows of interest will involve some amount of inflow and outflow from regions external to the domain. In a typical industrial example, the domain is an enclosed reactor with a number of inlets and outlets which are relatively small compared with the dimensions of the reactor. For the fluid dynamic balance equations, the number of boundary conditions necessary is determined by the number of incoming characteristics associated with the hyperbolic convection terms. Typically at an inlet, all characteristics are incoming so that the external conditions may be fully specified. At an outlet, however, generally all characteristics are outgoing, so that no boundary conditions should be set. When conditions are specified at the inlets and outlets, in general no continuous solution exists. However, the Boltzmann equation does not distinguish between inflow and outflow. The boundary condition for the Boltzmann equation is a velocity distribution function restricted to the domain $v_n = \vec{v} \cdot \vec{n} > 0$, where $\vec{n}$ is the normal pointing into the domain. This distribution function is weighted such that

$$M_{ent} = \int_S \int_{v_{n(x)} \geq 0} v_{n(x)} \, f(x, v) \, dv \, dx \qquad (2.64)$$

gives the total mass entering the domain per unit time through surface S. The distribution function f is almost always taken to be a Maxwellian. Therefore the stream velocity, density (or pressure), temperature and species composition must be specified for all inlets and outlets.

As an example we consider the case of steady flow into a two dimensional rectangular reactor. The inflow is taken to have a stream velocity $(u_x, 0, 0)$ with $u_x > 0$, so that the flow is in the positive x direction across a surface element with inward normal $(1, 0, 0)$. The distribution function of the incoming gas is taken to be a spatially homogeneous Maxwellian. The total flux per unit time per unit surface which should enter through the surface is then given by the formula

$$
\begin{aligned}
M_{ent} &= \frac{n}{(2\pi \frac{k_B}{m} T)^{3/2}} \int_{R^2} \int_{u_x}^{\infty} v_x \exp\left[-\frac{(v_x - u_x)^2 + v_y^2 + v_z^2}{2k_B T/m} \right] dv_x dv_y dv_z \\
&= n\sqrt{\frac{k_B T}{2m}} \left(s + s \, \mathrm{erf}(s) + \frac{1}{\sqrt{\pi}} e^{-s^2} \right) \qquad (2.65)
\end{aligned}
$$

where

$$s = \frac{u_x}{\sqrt{2k_B T/m}} \, .$$

For high Mach number space flight problems, these boundary conditions may be implemented without trouble. The flow is generally external to an object, and the domain is a box large enough to enclose the object and the flow disturbance which results from the objects presence. The up-stream boundary condition is a Maxwellian centered around the free stream velocity, while the downstream boundary is usually a vacuum. This approx-imation is based on the assumption that the free stream velocity is so large, that the tail of the velocity distribution function representing flow in to the domain in the opposite direction of the free streaming is negligible.

For reactor flow, the situation becomes more complicated. The stream velocities are generally considerably sub-sonic, so that the velocity distri-bution function at the outlet will have a significant back-flow tail. Thus it is necessary to specify conditions at the outlets, although these conditions may not be known a priori. Moreover, at the inlet, diffuse reflection at the reactor walls may produce significantly different temperatures and densities in the first cells than were given for the boundary distribution function. The result is that it may be necessary to iterate the boundary conditions to ob-tain continuous densities, temperatures and velocities which match (in the domain near the inlets) the desired quantities. While this is computationally expensive, usually only a couple of iterations are needed.

Another approach is to extend the domain to include regions outside the inlets and outlets. In this way the effect of the reactor on the incoming gas at the inlet may be computed. It is necessary to decide how the external flow (outside the domain of interest) should be modeled, for example as free flow or as flow through a pipe. In addition to the large domain, and corresponding larger computation time, the problem remains as to how to handle the inflow to the enlarged domain. As it would be extremely difficult to model all aspects of the flow before and after it reaches the reactor, it seems more practical to use the iterative approach to approximate the desired inlet conditions.

2.5 Free Molecular Flow

Free molecular flow refers to the limiting case when the gas is so rarefied that particle-particle collisions no longer play a significant role in determining the gas dynamics. These flows are characterized by a large Knudsen number, say $Kn > 10$. In this case the transport equation becomes simply

$$\frac{\partial f}{\partial t} + v \cdot \nabla_x f = 0 \,.$$

This is similar to radiation transport when there is no scattering in the bulk, but only surface interaction. Thus many of the numerical methods used in radiation transport are also used to compute free molecular flow.

Large Knudsen number flows arise frequently in microelectronics pro-cessing when microscopic device features of the wafer are considered. This is known as feature scale modeling. A typical problem is to model a trench

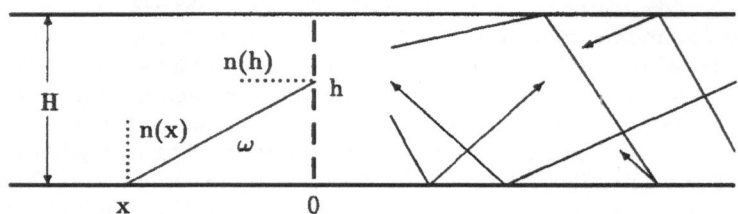

Figure 2.4: Knudsen flow through a 2D-tube.

filling process. Even at moderate pressures, the width of the trench may be many times smaller than the mean free path of the gas, so that within the trench free molecular flow dominates. In this case the gas-surface interaction completely determines the flow properties.

Under certain assumptions on geometry and surface scattering, the gas flux in the free molecular flow regime may be regarded as a diffusive flow. This is known as Knudsen diffusion. We derive now a formula for this flux and the corresponding diffusion coefficient for the case of a two dimensional flow through an infinitely long rectangular tube with diffusely reflecting walls. This situation is illustrated in Figure 2.4.

Following the development presented in Chapters 1 and 5 for radiation transport, the rate $q(x_0, h)$ at which molecules arrive at a certain point (x_0, h) inside the domain may be expressed as an integral over the surface of the domain

$$q(x_0, h) = \int_{-\infty}^{\infty} \Big(T[(x_0, h), (x, 0)]q(x, 0) + T[(x_0, h), (x, H)]q(x, H) \Big) dx \ .$$
(2.66)

The quantity $T[(x_0, h), (x, 0)]$ is a transition probability that a molecule at position $(x, 0)$ will reflect at the appropriate angle to arrive at the position (x_0, h). In the diffuse reflection approximation, T is called the view factor and may be shown to be

$$T[(x_0, h), (x, 0)] = \frac{hz}{(h^2 + z^2)^{3/2}}$$
(2.67)

where $z = |x - x_0|$.

The average flux density through a cross section of the tube at $x = 0$ can now be expressed solely in terms of impingement rate on the walls

$$j^{cross} = \frac{1}{H} \int_0^H q(0, h) dh$$
(2.68)

$$= \frac{1}{H} \int_0^H \int_{-\infty}^{\infty} \Big(T[(0,h),(x,0)]q(x,0) + T[(0,h),(x,H)]q(x,H) \Big) dx\,dh \ .$$

This may be simplified by using the view factor from Equation (2.67) and the symmetry of the domain with respect to h. The average flux is then expressed as

$$j^{cross} = \frac{2}{H} \int_0^H \int_{-\infty}^{\infty} \frac{hx}{(h^2+x^2)^{3/2}} q(x,0) dx\,dh \ . \qquad (2.69)$$

In the following the term $q(x,0)$ will be replaced by $q(x)$.

The geometry and the diffuse reflection at the walls suggest that the velocity distribution function in the interior of the domain should be Maxwellian. Then Equation (2.65) may be used to compute the molecular impingement rate on the wall. This is the special case for which $u_x = 0$ (or equivalently, $s = 0$), i.e., there is no stream velocity into the wall. In this case,

$$q = \frac{1}{4} n \, v_{ave} \ . \qquad (2.70)$$

Here n is the number density and v_{ave} is the average molecular speed, defined as

$$v_{ave} = (2\pi k_B/mT)^{-3/2} \int_{R^3} |\vec{v}| \exp\left(-\frac{|\vec{v}|^2}{2k_B/mT}\right) d\vec{v} = \sqrt{\frac{8k_BT}{\pi m}} \ . \qquad (2.71)$$

In order to have a net flux across a cross section of the tube, it is necessary to have a density gradient. It is reasonable to assume for a very long (infinite) tube that the density profile is in fact linear so that

$$q(x) = \frac{1}{4} v_{ave} \left(n(0) + \frac{\partial n}{\partial x} x \right) \ . \qquad (2.72)$$

The molecular flux then becomes

$$\begin{aligned} j^{cross} &= \frac{1}{4} v_{ave} \frac{2}{H} \int_0^H \int_{-\infty}^{\infty} \frac{hx}{(h^2+x^2)^{3/2}} \left(n(0) + \frac{\partial n}{\partial x} x \right) dx\,dh \\ &= H \, v_{ave} \frac{\partial n}{\partial x} \end{aligned} \qquad (2.73)$$

The quantity $D^{Knudsen} = H \, v_{ave}$ may be recognized as the Knudsen diffusion coefficient for a two dimensional rectangular tube. In the literature this is occasionally seen with an additional factor of $2/3$. It is assumed that this comes from using the cruder approximation that $q = 1/6 \, n \, v_{ave}$.

Chapter 3

Numerical Methods for Rarefied Gas Dynamics

Numerical methods for solving the gas dynamic Boltzmann equation are of key importance because in general analytic solutions can not be found. Only in special cases such as in the fluid dynamic or free molecular flow limits for specific geometries and boundary conditions are analytic solutions (or approximate solutions) known [15]. An interesting exception is the exact solution for the spatially homogeneous Boltzmann equation found by Krook and Wu [52]. However, for most situations of interest, the only means of solving the equation is computationally.

There are a number of numerical methods commonly employed for this task corresponding to the different flow regimes. By far the most developed are schemes for solving the Navier-Stokes equations which are appropriate in the hydrodynamic limit where the collisional relaxation time is much faster than the convection time scale so that the flow is very close to local equilibrium. A number of commercial software packages are available which are specifically designed for microelectronic applications [1], [2], [3]. Even in this familiar area, however, the question of boundary conditions for slightly rarefied flows has not been fully settled. The phenomenon of boundary slip in velocity, temperature and concentration is well known, but accurate modeling and determination of slip coefficients for gas mixtures remains to be done. This issue is discussed in the next chapter.

At the other extreme lies the free molecular flow regime in which collisions between particles are neglected and the flow is determined solely by particle interaction with the boundary. These problems are closely related to radiation transport, computer vision, and molecular deposition in microscopic features of microelectronic device structures, as well as other problems involving ray tracing. Depending on the boundary conditions, var-

77

ious methods are used including the view-factor method [47] for fully diffuse boundaries. For fully or partially specularly reflecting boundaries, typically a Monte Carlo simulation is used in which the main computational issues are efficient determination of where a particle strikes the boundary and what kind of collision occurs. These questions are addressed below in the discussion of the full Boltzmann equation simulation.

For flows described by the linear approximation to the Boltzmann equation, Monte Carlo simulations known as Test Particle methods have been developed [72]. These are typically used when a trace species is passing through a background gas. If it is assumed that the trace species particles only collide with the background and not among themselves, and that the background remains unchanged due to these collisions, then the background gas molecules need not be simulated at all, and the trace species may be treated as a sequence of independent test particles scattering in a fixed domain. A more detailed description of this process is given below, and the application of this method to a sputtering reactor is discussed in the next chapter.

Finally, flows in the transition regime, between the fluid dynamic and free molecular flow limits, must be considered. These flows are determined both by boundary interaction and particle-particle collisions, thus requiring the solution to the full Boltzmann equation. While quadrature methods [83] and discrete velocity methods [91] have been proposed to handle the integral term, the most successful and wide spread approach has been Monte Carlo simulation. This approach was pioneered by Bird [11] under the title Direct Simulation Monte Carlo (DSMC) beginning in the 1960s. Since then numerous developments, variations and interpretations have been introduced, a few of which can be found in [70], [51], [8]. Extensive testing on hypersonic space flight problems, as well as in industrial applications, has shown this non-linear particle simulation to be a computationally robust means of simulating rarefied flows. The method is detailed below.

The DSMC method is not limited to use in the transition regime; however it is extremely computationally expensive in the fluid dynamic limit. Given that there is emerging interest in industrial applications with flows with both fluid and transition regimes, there has been considerable interest in developing hybrid algorithms which couple Navier-Stokes solvers to DSMC algorithms. This involves establishing regions of the domain which can be considered fluid or transition, as well as computing the transport across an interface from one region to another. A number of research groups are currently working on this coupling.

3.1 Direct Simulation Monte Carlo

The original motivation for the development of the DSMC method was an attempt to simulate the physics of molecular motion and interaction without reference to the Boltzmann equation itself. It was not until the late 1980's that a convergence proof of Bird's method to a solution of the Boltzmann equation was given [96], although convergence for similar, though computationally less effective, methods derived directly from the Boltzmann equation were given a few years earlier [70], [7]. Numerous variations on the basic algorithm have been proposed, in particular with regard to parallel or vector computer architectures. For the sake of simplicity, only the basic Bird No Time Counter method will be described in detail.

We will begin with the simulation of a single species gas. DSMC is a particle simulation in which the particles represent gas molecules. From the mathematical, Boltzmann equation point of view, the particles are sampled points from the density distribution function $f(x, v, t)$. If at time t the simulation points have positions and velocities $\{(x_i, v_i), 1 \leq i \leq N\}$, then the distribution function is approximated as

$$f(x, v, t) \approx \frac{M}{N} \sum_{i=1}^{N} \delta(x - x_i)\, \delta(v - v_i) \tag{3.1}$$

where M is the total mass in the domain and δ is the Dirac measure. The simulation updates $\{(x_i, v_i)\}$ as specified by the Boltzmann equation.

A more physical interpretation is that each simulation particle represents a large number of real gas molecules, and thus the dynamics of the particles should be the same as those of a molecule with the same velocity. On the individual particle level, the mathematical and physical interpretations are quite similar because the Boltzmann equation is derived from considering two particle interactions. The difference lies in the fact that the Boltzmann equation is derived as a limit of the number of particles goes to infinity, whereas the physical simulation approach is more like solving a master equation for a finite system. The algorithm described here may be obtained through either approach. It is presented here in the Boltzmann equation framework.

The goal of the method is to compute the development in time of the distribution function $f(x, v, t)$ which depends on both space and velocity. The physical, spatial domain is typically three dimensional axi-symmetric (thus effectively two dimensional) or fully three dimensional. Occasionally two dimensional rectangular problems are of interest, such as flow over a flat plate of infinite width. One dimensional problems such as planar shock waves or heat transfer between flat plates are often used as test problems or to compare computed parameters to experimentally measured values.

Velocity space must remain unbounded and is almost always taken to be fully three dimensional, even when it might be reduced for one and two dimensional spatial domains. Due to the nature of the method, only very limited computational savings could be gained by such a reduction. Usually the simplicity gained by maintaining consistency across dimensions outweighs the potential improvement in computation time.

For most applications, relaxation to a steady state occurs over a fast time scale relative to the time periods of interest. Thus in general, only steady solutions are sought. Numerically, however, the steady solution is obtained by solving the time dependent problem starting from arbitrary initial data and running until the steady state is reached. An intelligent initial guess for the flow field will speed convergence to the steady state. Because the method is statistical in nature, once the steady state has been reached the algorithm continues to run until an adequate sample has been collected. Should an unsteady solution be required, the same algorithm may be applied. In this case, however, the results from a number of different, independent runs of the algorithm (produced by rerunning the code with different sequences of random numbers) are averaged together to reduce the statistical fluctuations.

3.1.1 Spatial Discretization

Cell Size

The spatial domain is divided into cells. Depending on geometry, the cells may be part of a rectangular lattice or may be determined by a body-fitted coordinate system. A characteristic cell dimension Δx is chosen such that Δx is some fraction of a mean free path (typically $1/3$ to $1/2$). This constraint on cell size illustrates two important aspects of the algorithm. First, as mean free path decreases, for example as the gas becomes more dense, the cell size decreases, so that the number of cells required increases rapidly. This shows why flows near the fluid dynamic limit can be quite expensive to compute. Second, the mean free path may vary considerably across the domain so that a variable mesh size may be essential for efficient computation. Adaptive grid generation and domain decomposition techniques are often employed in these cases. This tends to be of greater concern in supersonic space flight than in industrial applications, where the range of temperatures and spatial scales is more restricted.

In the following, it will be assumed that a uniform rectangular grid is adequate. The case of axi-symmetric grids is considered later. For the purpose of computing the mean free path, the gas will initially be taken to be a single species; the extension to multiple species will be discussed subsequently. An initialization routine establishes the domain and the initial conditions. This is easily done by specifying the size of the domain and the number of cells (or the cell size) and the initial distribution function within each cell. Unless some a priori information is available (which is usually not the case), the distribution functions are taken to be spatially homogeneous (across each cell) Maxwellian distributions, so that only density, temperature and mean velocity need be specified in each cell. A typical mean free path is calculated as a function of density and temperature. The cell size is then compared to the mean free path to ensure that the size constraint has been met. The final results tend to be relatively insensitive to the cell size; however, it is wise to check that the mean free path of the final answer is indeed larger than the cell size used to compute it.

Mean Free Path

The mean free path λ is calculated from the formula

$$\lambda = c \frac{\mu}{\rho\sqrt{k_B T/m}} \tag{3.2}$$

which follows from the Chapman-Engskog expansion for the Boltzmann equation. Here the constant c and the viscosity $\mu = \mu(T)$ depend on the collision law chosen. Again, k_B is the Boltzmann constant, m is the molecular mass, and ρ and T are the density and temperature.

As discussed in the previous chapters, for inverse power law molecular potentials, the Chapman-Engskog expansion leads to a dependence of viscosity on temperature of the form

$$\mu(T) = \mu_{\text{ref}} \left(\frac{T}{T_{\text{ref}}}\right)^\omega \tag{3.3}$$

where μ_{ref} is a reference viscosity measured at temperature T_{ref}, and $0.5 \leq \omega \leq 1$. The most widely used collision model for DSMC calculations, the Variable Hard Sphere (VHS) model (see Section 2.4.1) introduced by Bird [13], [14], is based on this viscosity law. For the VHS model, the constant c in Equation (3.2) is

$$c = \frac{2(7 - 2\omega)(5 - 2\omega)}{15\sqrt{2\pi}} . \tag{3.4}$$

Thus the mean free path of the gas may be determined by specifying the molecular mass, a reference viscosity and temperature, a viscosity exponent ω, and the density and temperature of the gas.

3.1.2 Time Step

The Boltzmann equation may be written as

$$\frac{\partial f}{\partial t} = -\mathcal{C}f + \mathcal{Q}f \tag{3.5}$$

where C is the convection operator

$$Cf = v \cdot \nabla_x f \qquad (3.6)$$

and Q is the collision operator

$$Qf = \frac{1}{m} Q(f, f) . \qquad (3.7)$$

The time integration is carried out with a simple forward Euler method over the time step Δt. Formally this may be written

$$f^{n+1} = \left(1 - \Delta t \, C + \Delta t \, Q \right) f^n + \mathcal{O}(\Delta t^2) \qquad (3.8)$$

where $f^n = f(x, v, t^n)$ is the solution at time $t^n = n\Delta t$. To the same degree of accuracy, the operator may be factored as

$$f^{n+1} = \left(1 - \Delta t \, C \right) \left(1 + \Delta t \, Q \right) f^n + \mathcal{O}(\Delta t^2) . \qquad (3.9)$$

The Monte Carlo simulation proceeds under the assumption that the time step is chosen small enough that this decoupling of the collision and convection steps is valid. This may be ensured by choosing the time step such that a particle traveling at the macroscopic flow speed plus the most probable thermal speed,

$$v = |v_0| + \sqrt{2\frac{k_B}{m}T} , \qquad (3.10)$$

travels less than one cell per time step. The time step is then set in the initialization routine by using v_0 and T taken from typical values of the initial data. As with cell size, the time step should be checked against the final result to ensure that this condition remains satisfied.

This restriction on time step is similar to the Courant Friedrich Lewy (CFL) condition for the compressible Euler equations, which states that

$$\frac{\Delta x}{\Delta t} < |v_0| + c , \qquad (3.11)$$

where c is the sound speed of the gas, which is roughly the same as the most probable thermal speed. The time step restriction is also equivalent to requiring that the time step be smaller than a mean collision time.

3.1.3   Collision Step

The decoupling of the time discretized Boltzmann equation given in Equation (3.9) means that collisions and convection may be treated separately. The algorithm consists of alternating between computing collisions and convecting the flow.

The collision operator $Q(f, f)$ acts only on velocity space, so that the collision step means solving the spatially homogeneous Boltzmann equation. Using the time discretization previously described, this may be written

$$f^{n+1}(v) = \left(1 - \frac{\Delta t}{m} L(f^n, v)\right) f^n(v) + \frac{\Delta t}{m} G(f^n, f^n, v). \qquad (3.12)$$

The loss operator L is defined as

$$L(f, v) = \int_{\mathcal{R}^3} \int_{S+} f(w)B(|v - w|, \omega)d\omega \, dw, \qquad (3.13)$$

and the gain operator G is

$$G(f, f, v) = \int_{\mathcal{R}^3} \int_{S+} f(v')f(w')B(|v - w|, \omega)d\omega \, dw, \qquad (3.14)$$

where v' and w' are defined in Equation (1.52).

Equation (3.12) may be interpreted probabilistically as saying that the probability that a molecule with velocity v does not collide in time step Δt is $1 - \frac{\Delta t}{m} L(f^n, v)$. The probability that a molecule with velocity v does collide in time step Δt is therefore

$$P(v) = \frac{\Delta t}{m} L(f^n, v) \qquad (3.15)$$

This appears to put another constraint on the time step, requiring that it be chosen small enough such that $P(v) \leq 1$ (and the corresponding probability of not colliding is non-negative). Conceptually this problem may be avoided by dividing the convection time step Δt into n_t suitable small "collision time steps" $\Delta \tau$ and repeating the collision algorithm using $\Delta \tau \, n_t$ times. In practise this simply allows a simulation particle to collide more than once during the collision step. In some vectorized algorithms this idea of the sub-time step is implemented; however, it is not necessary if $P(v)$ is interpreted as the number of times (on average) a simulation particle with velocity v should collide in time Δt.

At this point the necessity of cell structure described above may be explained. In order to perform the collision step described by Equation (3.12), a representation for the distribution function f is required. A particle approximation such as in Equation (3.1) is used. However, it is clear that with a finite number of particles it is not possible to have a particle representation of f at every spatial point x. Hence the delta functions in x as well as v in Equation (3.1). In order to construct the spatially homogeneous distribution f for the collision step from the $f(x, v, t)$ given by Equation (3.1), a cell structure is imposed on the domain in which the cells are chosen small enough that the true distribution function may be considered spatially homogeneous across each cell. A new distribution function $\tilde{f}$ for the cell is defined as the integral of the original, spatially inhomogeneous distribution function over the cell

$$\tilde{f}(v, t) = \int_C f(x, v, t)dx .$$ (3.16)

Then each particle in a cell is simply one point sampled from the spatially homogeneous distribution for that cell. In this way a $\tilde{f}(v)$ may be constructed for each cell from the particles with x coordinates in that cell. The assumption of spatial homogeneity means that for the purposes of the collision step, the actual position of a particle is irrelevant, other than for determining in which cell it resides. For a cell in which there are N particles,

$$\tilde{f} = \frac{mnV_c}{N} \sum_{i=1}^{N} \delta(v - v_i)$$ (3.17)

where v_i refers to the velocity of the ith particle in the cell. The quantity n is now the number density of real gas molecules in the cell, which has volume V_c, such that mnV_c is the total mass in the cell. The number density n must be related to the number of simulation particles N in the cell. This is done through the formula

$$nV_c = N \left(\frac{n_0 V_c}{N_0} \right) .$$ (3.18)

Here n_0 is the number density of the initial data in the cell and N_0 is the original number of simulation particles placed in the cell at the beginning of the simulation. The quantity in parentheses can be interpreted as the number of real gas molecules represented by each simulation particle. This is set in the initialization routine. If the grid is such that all cells have equal volume, then this value will be constant for all cells. If this constant is called p, then the last equation becomes

$$nV_c = N p .$$ (3.19)

The probability $P_i = P(v_i)$ can now be computed as

$$P_i = p\Delta t \sum_{j=1}^{N} (\sigma V_r)_{ij}$$ (3.20)

where

$$(\sigma V_r)_{ij} = \int_{S+} B(|v_i - v_j|, \omega)d\omega .$$ (3.21)

Here σ indicates the total collision cross section and V_r is the relative velocity of the colliding particles

$$(V_r)_{ij} = |v_i - v_j| .$$ (3.22)

For hard sphere collisions, $\sigma = \pi d^2$ is a constant where d is the particle diameter. In general, σ is a function of the relative velocity.

From Equation (3.20) the probability (or perhaps better, the number of times on average) that simulation particle i collides with simulation particle j is

$$P_{ij} = p\Delta t\,(\sigma\,V_r)_{ij}\,. \qquad (3.23)$$

If $(\sigma\,V_r)_{\max}$ is a constant upper bound chosen so that

$$(\sigma\,V_r)_{\max} \geq (\sigma\,V_r)_{ij} \qquad (3.24)$$

for all pairs of simulation particles i and j, then Equation (3.23) may be rewritten as

$$P_{ij} = \left[\frac{1}{2}N(N-1)p(\sigma V_r)_{\max}\Delt\right]\left[\frac{2}{N(N-1)}\right]\left[\frac{(\sigma V_r)_{ij}}{(\sigma V_r)_{\max}}\right]\,. \qquad (3.25)$$

No Time Counter method

This name is derived from the original Bird collision algorithm [11] which was called the Time Counter method. That method involved computing a collisional time associated with each collision and performing collisions until the sum of all the collisional times exceeded the convection time step. It was found that if a highly unlikely collision occurred, the associated collisional time would be large enough to effectively prohibit any further collisions in the cell for a number of subsequent convection time steps. To avoid this problem, the No Time Counter (NTC) method was introduced.

The NTC method is based on Equation (3.25). The idea is to select a pair of particles from the list of particles in the cell and, with a specified probability, accept or reject that pair as collision partners. If the pair is accepted, then new velocities for the pair are computed according to the collision law. These velocities replace the pre-collision velocities. Other factors such as rotational and vibrational energies and chemical reactions may be computed as well, if those effects are being taken into account. If the pair is rejected, then the particles retain their precollision velocities (and energies, etc.). Equation (3.25) indicates how many such pairs should be selected, how they should be selected and the probability of accepting the collision pair in the following way. The number of potential collision pairs (i.e., the number of pairs to be selected for testing) N_c is

$$N_c = \frac{1}{2}N(N-1)p(\sigma V_r)_{\max}\Delta t\,. \qquad (3.26)$$

The quantity $2/(N(N-1))$ is the probability of uniformly selecting any pair from the set of all possible particle pairings. Finally, $(\sigma V_r)_{ij}/(\sigma V_r)_{\max}$ gives the probability that the pair (i,j), having been selected, will collide.

A number of remarks on the method should be made at this point. From Equation (3.26) it appears that the number of collisions to be computed is $\mathcal{O}(N^2)$. However, it must be remembered that p is inversely proportional to N_0, and is therefore inversely proportional to N. Thus the number of collision which must be computed is in fact only $\mathcal{O}(N)$. This is of key importance because if the method were quadratic in N, even simple problems might require days to compute, and complicated three dimensional flows would be all but intractable.

This linearity in N is perhaps initially surprising considering the quadratic nature of the collision operator Q. It might seem that computing the probability that particle i collides would require computing the sum in Equation (3.20). When this is done for all particles, the method becomes $\mathcal{O}(N^2)$. This is avoided by introducing $(\sigma V_r)_{\max}$ into the calculation. This allows for the sum in Equation (3.20) to be evaluated with the Monte Carlo acceptance-rejection technique described above.

In order to use this method, it is necessary to assign a value to $(\sigma V_r)_{\max}$. From Equation (3.26) it is clear that the value should be as small as possible so as to minimize the number of collision pairs which must be selected, but large enough to bound all the relative velocities in the cell. The exact maximum value of $(\sigma V_r)_{ij}$ could be computed in $\mathcal{O}(N^2)$ operations, but this is to be avoided. It is possible to obtain an estimate of the upper bound in $\mathcal{O}(N)$ operations which is generally within 10% of the exact value. However, the exact upper bound changes after every collision, so that this upper bound would also have to be recomputed after every collision, which becomes expensive. The practical solution proposed by Bird is to set an initial estimate of $(\sigma V_r)_{\max}$ for each cell in the initialization routine, and simply update this value should a collision pair be encountered for which $(\sigma V_r)_{ij} > (\sigma V_r)_{\max}$ by setting $(\sigma V_r)_{\max} = (\sigma V_r)_{ij}$. After a short transient, an accurate estimate for the maximum is usually obtained.

VHS and M1 Models

It remains to be said how the collision cross section $(\sigma V_r)_{ij}$ defined in Equation (3.21) is evaluated. The standard models for the intermolecular potential are the inverse power law described in Chapter 1, or the related Lennard-Jones potential. As discussed in Section 2.4.1 the difficulty in implementing these collision models is that the scattering angle χ defined in Equation (2.40) is given as an integral which must be evaluated for every collision. The practical alternative is to model the cross section with the Variable Hard Sphere (VHS) model of Bird [13] or the related M1 model discussed in the last chapter.

If the VHS model is used, the $(\sigma V_r)_{ij}$ need for the NTC method is then simply calculated as

$$(\sigma V_r)_{ij} = \pi d_{VHS}^2(V_r)V_r \tag{3.27}$$

where $V_r = |v_r| = |v_i - v_j|$ is the relative velocity of the colliding particles i and j, and $d_{VHS}(V_r)$ is given by Equation (2.52). If the pair is accepted to collide, the post-collision velocities are computed based on a hard sphere collision model. In the center of mass reference frame this corresponds to the post-collision scattering direction being uniformly distributed over the unit sphere. The post-collision velocities v_i' and v_j' may be expressed as (see Section 1.2)

$$v_i' = (v_m)_{ij} + \frac{m_r}{m_j} V_r \vec{n}' \tag{3.28}$$

$$v_j' = (v_m)_{ij} - \frac{m_r}{m_i} V_r \vec{n}' .$$

Here m_r is the reduced mass, v_m is the center of mass velocity

$$(v_m)_{ij} = \frac{1}{2}(v_i + v_j) , \tag{3.29}$$

and $\vec{n}'$ is a unit vector sampled uniformly over the sphere. This sampling is done using two uniformly distributed random numbers ξ_1, ξ_2 on the interval [0,1] and setting $\vec{n}' = (n_x', n_y', n_z')$ with

$$n_x' = 1 - 2\xi_1 \tag{3.30}$$

$$n_y' = 2\sqrt{\xi_1(1 - \xi_1)} \cos(2\pi\xi_2)$$

$$n_z' = 2\sqrt{\xi_1(1 - \xi_1)} \sin(2\pi\xi_2)$$

For more general scattering angle dependencies than the hard sphere model, more care must be taken in the computation of $\vec{n}'$. Typically, an impact parameter b should be sampled uniformly on the range $[0, b_{max}]$. From this the scattering angle $\chi(b)$ may be computed. The rotation angle ψ is taken to be uniformly distributed, so that $\psi = 2\pi\xi$ where ξ is a uniform random number selected from the interval [0,1]. From χ and ψ, the vector $\vec{q}$ may be computed

$$\vec{q} = (\cos\chi, \ \sin\chi \ \cos\psi, \ \sin\chi \ \sin\psi) . \tag{3.31}$$

As discussed in Section 1.2, this vector gives the post-collision direction of the relative velocity in the reference frame in which the coordinates of the pre-collision relative velocity direction $\vec{n}$ are $(1, 0, 0)$. As in Equation (1.11), the vector $\vec{q}$ must be multiplied by the orthonormal rotation matrix $R(\vec{n})$. The matrix can be explicitly written as

$$R(\vec{n}) = \left[\ \vec{n} \ \ \vec{n}_1^\perp \ \ \vec{n}_2^\perp \ \right] \tag{3.32}$$

where $\vec{n} = (n_x, n_y, n_z)$ is the direction of the pre-collision relative velocity and $\vec{n}_1^\perp$ and $\vec{n}_2^\perp$ are given by

$$\vec{n}_1^\perp = \frac{1}{\sqrt{n_y^2 + n_z^2}} \begin{pmatrix} 0 \\ n_z \\ -n_y \end{pmatrix} , \quad \vec{n}_2^\perp = \frac{1}{\sqrt{n_y^2 + n_z^2}} \begin{pmatrix} n_y^2 + n_z^2 \\ -n_x n_y \\ -n_x n_z \end{pmatrix} \quad (3.33)$$

It follows that the post-collision relative velocity direction is given by

$$\vec{n}' = \cos\chi \; \vec{n} + \sin\chi \; \cos\psi \; \vec{n}_1^\perp + \sin\chi \; \sin\psi \; \vec{n}_2^\perp . \quad (3.34)$$

More explicitly, this is

$$n_x' = \cos\chi \; n_x + \sin\chi \; \sin\psi \; \sqrt{n_y^2 + n_z^2} \quad (3.35)$$

$$n_y' = \cos\chi \; n_x + \sin\chi \; \frac{n_z \; \cos\psi - n_x n_y \; \sin\psi}{\sqrt{n_y^2 + n_z^2}}$$

$$n_z' = \cos\chi \; n_x - \sin\chi \; \frac{n_y \; \cos\psi + n_x n_z \; \sin\psi}{\sqrt{n_y^2 + n_z^2}}$$

In the case of the M1 model, $(\sigma V_r)_{ij}$ is computed from Equation (3.27), just as with the VHS model. However, now the scattering is not uniformly sampled from the unit sphere. The scattering angle χ and rotation angle ψ are computed as

$$\chi = \pi \xi_1 \quad (3.36)$$
$$\psi = 2\pi \xi_2$$

where again ξ_1 and ξ_2 are uniformly distributed random numbers on the interval $[0,1]$. The above formulas may be used to determine the post-collision velocities.

3.1.4   Convection Step

The convection step of the DSMC algorithm is conceptually simpler than the collision step. The operator splitting of Equation (3.9) indicates it involves solving the free flow equation

$$\frac{\partial f}{\partial t} + v \cdot \nabla_x f = 0 . \quad (3.37)$$

This equation simply states that a particle travels on a straight path (in the absence of external fields) in the direction of its velocity v. The linearity of the equation indicates that there is no interaction among the particles. Thus given the initial positions and velocities of the particles, it is a matter of ray tracing to determine the final positions after a time step Δt. In the absence of boundaries, the solution at time $t + \Delta t$, given initial data $f(x, v, t)$ from Equation (3.1), is

$$f(x, v, t + \Delta t) = \frac{M}{N} \sum_{i=1}^{N} \delta(v - v_i)\delta(x - (x_i + v_i \Delta t)), \qquad (3.38)$$

which means that

$$x_i^{n+1} = x_i^n + v_i^n \Delta t \qquad (3.39)$$

and

$$v_i^{n+1} = v_i^n . \qquad (3.40)$$

The superscript n indicates position and velocity at time t^n.

3.1.5 Boundaries

The presence of boundaries complicates matters somewhat. Details of deciding exactly where and when a boundary collision occur are discussed below. First, however, the procedure for computing the interaction of a particle with a surface is presented.

Wall Interaction

Equations (2.62) and (2.63) give the formulas for specular and diffuse reflection at a wall. These may be used together with a thermal accommodation coefficient α to more realistically approximate surface interaction. In the simulation of partial thermal accommodation, ξ, a uniform random number on [0,1], is generated for each surface collision. If $\xi < \alpha$, the particle is diffusely reflected. Otherwise, specular reflection is used. If diffuse reflection is selected, then the new particle velocity is sampled from the distribution given in Equation (2.63). This may be achieved by using a rejection technique. For specular reflection, the new velocity is computed deterministically.

The time required to travel from the original particle position x_i to x_w, the point of impact with the surface, is

$$\Delta t_c = \frac{|x_w - x_i|}{|v_i|} . \qquad (3.41)$$

Necessarily, $\Delta t_c < \Delta t$. Therefore the particle must continue to travel for the rest of the convection time step $(\Delta t - \Delta t_c)$ with the new, post-reflection velocity. This assumes that the collision process is instantaneous. The particle may continue on to another (or several other) collision(s) before the convection time step is finished.

While the conceptual formulation of the free flow is simpler, the convection step may be the most difficult part of the algorithm to code and the most expensive to compute. Complicated domains may have many surfaces, and determining precisely which surface and where a particle strikes requires some effort.

One approach is to move particles one cell at a time. Starting from the interior of a cell, a particle is first moved to the edge of the cell and then moved across the adjacent cell. Given that the time step has been chosen so that particles will generally move less than once cell per time step, not more than one or two steps will be required for most particles. Boundary cells may be specially marked so that when a specified edge is reached, the particle reflects instead of continuing on.

If the domain contains large regions without boundaries, it may be more efficient to divide the domain into macroscopic cells as well as the microscopic cells used in the collision algorithm. The idea is to decompose the entire position space into disjoint convex sets which are determined by the boundaries of the domain. These sets will include regions which may be inaccessible to particles (because of walls), and the exterior sets will be unbounded. In a rectangular geometry, the macro-cells will be aligned with the micro-cell grid structure so that integer arithmetic may be used to determine the cell number (both macro and micro) from the position coordinate x.

Macro-cells are used to determine whether a particle has crossed a boundary. First, free flow without boundaries is performed according to Equation (3.39). The macro-cell containing the final position is compared with the macro-cell of the initial position. If the initial and final positions are within the same macro-cell, then the convexity of the cell assures that that the entire trajectory was contained in the cell, so that no boundary collision arises. Associated with each pair of macro-cells is a list of all surfaces which could be crossed in connecting the two cells. If the initial and final cells are different, then the list of possible surfaces is checked, and the surface crossed by the trajectory, if any, is determined. If the macro-cells are large compared to the micro-cells, then the vast majority of particle trajectories will lie within the same macro cell and no surface interaction need be checked.

The two methods may be combined such that particles starting in certain regions follow the macro-cell approach, while particles in other regions (close to complicated boundaries) follow the cell to cell approach.

Inflow and Outflow

The procedure for introducing new particles in the domain due to influx is similar to diffuse reflection. It is assumed that the distribution function at the inlet or outlet is a Maxwellian, so the velocities of the incoming particles are sampled from a flux weighted Maxwellian in the direction of the surface normal. The mass influx per unit time M_{ent} through a surface S is given by Equation (2.64) in Chapter 2. Typically an inlet or outlet will border a number of cells in the domain. To minimize statistical fluctuations, the flux should be computed for each cell based on the external Maxwellian chosen for that cell and the surface area of the cell. The number of particles entering the cell during a time step Δt is given by

$$N_{ent} = \frac{1}{m} M_{ent} \Delta t \qquad (3.42)$$

The particles are assigned entry times which may be chosen randomly, as in a Poisson process. However, fewer fluctuations are introduce if the random event time of this process (the time between particles entering) is replaced by its expectation, which in this case is

$$\Delta t_{ent} = \frac{\Delta t}{N_{ent}} . \qquad (3.43)$$

The quantity N_{ent} will not be an integer, and should not be rounded to an integer. Instead, it is used to compute Δt_{ent}. For each inflow cell a running time counter is kept which is incremented by Δt_{ent} for every particle introduced. The sum is taken modulo Δt, and the particle which causes the sum to exceed Δt is held back and not introduced until the next convection time step.

3.2 Computing Results

The simulation procedure just described computes a time dependent flow which should tend to a steady state flow. The number of time steps required to reach the steady state may vary considerably, depending on whether the flow is convection or diffusion driven, the density of the gas, the closeness of the initial data to the steady solution, etc. Precautions should be taken to ensure that the steady state has been reached before the results are computed.

Once the flow has become steady there will still be large fluctuations in each cell from time step to time step. This is because only a relatively small number of simulation particles (perhaps 10 to 100) will be present in a cell at any given time. The particle velocities are considered to be sampled from the true velocity distribution function. Because the flow is steady, the true distribution function will not change with time, so at the next time step a new set of sampled velocities may be collected. In this way it is possible to collect a large sampling of the distribution function over a many time steps, and thus obtain an accurate representation of the solution to the Boltzmann equation.

For the most part, the actual velocity distribution function solution of the steady Boltzmann equation is not the quantity of interest. The true goal of the calculation is to compute various moments of the distribution function which represent the macroscopic mass density, stream velocity, temperature, etc. The quantities are obtained from various functions of the particle velocity being integrated against the velocity distribution. The particle representation of the distribution function provides a natural means of evaluating these integrals Monte Carlo style. If in a given cell a total sample size of N_{tot} particles is collected over N_{time} time steps, p is defined in Equation (3.19) as the number of real gas molecules represented by a simulation particle, and m is the molecular mass, then the mass density is approximated by

$$\rho = \int_{\mathbb{R}^3} f(v)dv \approx \frac{mpN_{tot}}{N_{time}} .\qquad(3.44)$$

The stream velocity of the gas in the cell is approximated as

$$v_0 = \frac{1}{\rho} \int_{\mathbb{R}^3} vf(v)dv \approx \frac{1}{N_{tot}} \sum_{i=1}^{N_{tot}} v_i . \qquad(3.45)$$

If k_B is the Boltzmann constant, then the temperature T is defined through

$$3\frac{k_B}{m}T = \frac{1}{\rho} \int_{\mathbb{R}^3} |v - v_0|^2 f(v)dv . \qquad(3.46)$$

To approximate this correctly it is important to remember that v_0, which is constant as far as the integration is concerned, is also obtained as an approximation. This eliminates one of the N_{tot} degrees of freedom, so the approximation is

$$3\frac{k_B}{m}T \approx \frac{1}{N_{tot}-1} \sum_{i=1}^{N_{tot}} |v_i - v_0|^2 \qquad(3.47)$$

$$= \frac{N_{tot}}{N_{tot}-1} \left(\frac{1}{N_{tot}} \sum_{i=1}^{N_{tot}} |v_i|^2 - |v_0|^2 \right) .$$

The second form of the approximation is most useful because only the running totals of the sums of velocity and square of velocity need to be stored at each time step, as opposed to storing the velocity of every particle that has appeared in the cell.

At the density computed here most gases are assumed to satisfy the ideal gas law equation of state so that pressure is computed simply as

$$P = \rho \frac{k_B}{m} T . \qquad(3.48)$$

The heat flux, defined as

$$q = \frac{1}{2} \int_{\mathbf{R}^3} (v - v_0)|v - v_0|^2 f(v) dv \, , \qquad (3.49)$$

may be also be computed in a similar fashion to temperature.

An interesting question is whether it is better to use more particles per cell and average over fewer time steps, or use fewer particles with more averaging. Apart from maintaining at least a few particles in each cell, the difference in the computational results is relatively small. Depending on the computer, there may be some computational advantage for a certain range of particles. Typical computations use anywhere from 10 to a few hundred particles per cell. Availability of memory may be the determining factor in setting the upper limit.

3.3 Extensions

3.3.1 Multiple Species

When the flow of a multi-species rarefied gas is considered, the coupled set of Boltzmann equations described in Chapter 1 must be solved. The numerical algorithm for a multi-species gas is however essential the same as the single species version. The convection step remains unchanged except that each species may have a different accommodation coefficient at the walls, and a mass fraction for each species must be given at the inlets and outlets.

The collision algorithm is also virtually unchanged. The number of possible collisions occurring in a cell is still computed based on the number of particles in the cell and an upper bound on all the collision cross sections. The difference is when particles i and j are selected as potential collision partners, their collision cross section is computed as

$$(\sigma V_r)_{ij} = \frac{\pi}{4} \left(d_{i,VHS} + d_{j,VHS} \right)^2 V_r \qquad (3.50)$$

where $d_{i,VHS}$ is the VHS diameter (depending on relative velocity) of the species to which particle i belongs as given in Equation (2.52).

Finally, the definition of the mean free path, used to determine cell size, needs to be modified. For a gas of N_s different species, if n_i is the density of species i, and n is the total density, then the species-averaged mean free path is

$$\lambda = \sum_{i=1}^{N_s} \frac{n_i}{n} \lambda_i \, . \qquad (3.51)$$

Here λ_i is the mean free path of species i defined by

$$\frac{1}{\lambda_i} = \sum_{k=1}^{N_s} \frac{\pi}{4} \left(\bar{d}_{i,VHS} + \bar{d}_{k,VHS} \right)^2 n_k \left(1 + \frac{m_i}{m_k} \right)^{\frac{1}{2}} . \tag{3.52}$$

$\bar{d}_{i,VHS}$ is the VHS diameter of species i evaluated at the root mean square relative velocity for Maxwellian distributions of species i and k at temperature T. This relative velocity is given by

$$\overline{V}_r = \sqrt{\frac{3k_B T}{m_r}} \tag{3.53}$$

where m_r is the reduced mass of species i and k.

3.3.2 Axisymmetric Flow

More realistic three dimensional flows may be computed with the same effort as two dimensional rectangular flows under the assumption of rotational symmetry about an axis. The algorithm is basically the same as the 2-D rectangular case, and the collision step is identical. There are two major differences, though, which should be clarified.

First, for axisymmetric flows there is an option in choosing a uniform grid. It is assumed that the cells are of equal length is the direction of the axis of symmetry. In the radial direction, however, the cells may be chosen to have uniform length, in which case the volume of the annulus which each cell represents increases quadratically; or the cells may be chosen to have uniform volume, in which case the radial length of the cells decreases like a square root as radius increases.

The advantage of cells of uniform radial length is that the number of cells increases only linearly as the radius increases. However, the number of particles in each cell increases with radius (corresponding to the larger volume); for example, if N_C cells are used for a uniform density calculation, there will be approximately $2N_C$ times as many particles in the outer cell as in the cell at the axis. Thus there will be larger statistical fluctuations in the interior cells than in the outer cells.

Equal volume cells have the advantage that each cell will have a number of particles proportional to its density. However, the number of cells will increase quadratically with radius. This leads to longer computation times. Consider the constant density case in a domain of radius R and compare the two approaches. The inner most cell will be taken to be the same size in each case, with radius r. From the discussion of the no time counter method above, it can be seen that the number of collision per cell is proportional to $N\mathcal{V}_c$, where N is the number of particles in the cell, and $\mathcal{V}_c$ is the volume of the cell. Because N is proportional to $\mathcal{V}_c$, the number of collisions is in fact proportional to $\mathcal{V}_c^2$. The total number of collisions in a time step is therefore proportional to the sum over all radial cells of $\mathcal{V}_c^2$. In the equal volume case, this is approximately equal to $\mathcal{V}_1^2 N_{C,vol}^2/2$, where $N_{C,vol}$ is the number of equal volume cells and $\mathcal{V}_1$ is the volume of the first cell (the same for both methods). In the equal radial length case, the total number of collisions is proportional to $4\mathcal{V}_1^2 N_{C,rad}^3/3$, where $N_{C,rad}$ is the number of equal radius cells. But $N_{C,rad} = R/r$ while $N_{C,vol} = (R/r)^2$, so that the equal volume method requires $3R/8r$ times as many collisions per time step. The accuracy will be the same as the lowest accuracy equal radius cell (that is, the first cell). However, for the cost of the extra collisions (proportional to R/r), one obtains $(R/r)^2$ data points, as opposed to R/r in the equal radial length case. On the other hand, if the interior most cell is chosen small enough to satisfy the spatial homogeneity assumptions, then these extra points will probably be providing unnecessary refinement. This argument suggests that unless additional spatial refinement is required near the outer edge of the cylinder, it is more efficient to use the equal radial length approach.

The second difference for cylindrical geometry is the necessity of rotating the velocities after a convection step back into the radial and tangential reference frame. The cylindrical spatial coordinates are (r, θ, z). Under cylindrical symmetry, all particles are initially placed in the $\theta = 0$ plane. However, the particles have three dimensional velocities given by Cartesian coordinates (u, v, w). Taking the $\theta = 0$ plane to be the spatial Cartesian half-plane $(x, 0, z)$, $x \geq 0$, then a particle will initially have Cartesian coordinates $(r_0, 0, z_0)$. After a time step Δt of free convection, its position is $(r_0 + u\Delta t, v\Delta t, z_0 + w\Delta t)$. The new radial position is

$$r' = \sqrt{(|r_0 + u\Delta t|^2 + |v\Delta t|^2}. \tag{3.54}$$

It is necessary to rotate this vector back into the $\theta = 0$ plane. This is done by applying the rotational transformation

$$Q = \begin{bmatrix} \cos\phi & \sin\phi \\ -\sin\phi & \cos\phi \end{bmatrix} \tag{3.55}$$

where

$$\sin\phi = \frac{v\Delta t}{r'}. \tag{3.56}$$

The result of this transformation applied to $(r_0 + u\Delta t, v\Delta t)$ is $(r', 0)$, as expected. The point, though, is that the Cartesian velocities (u, v) must also be rotated so that $(u', v') = Q(u, v)$. Thus u' is the radial velocity of the particle and v' is the velocity in the θ direction at $\theta = 0$. The velocity w in the z direction remains unchanged. Reflections from walls should be handled in a similar manner.

3.3.3 Gas Phase Chemical Reactions

A useful approach for including gas phase chemical reactions in DSMC calculations has been developed for supersonic flows [12]. As gas phase reactions may be important in semiconductor processing at reduced pressure, a brief summary of the technique is presented here.

The inclusion of gas chemistry modifies only the collision step of the algorithm, which is necessarily now for multi-species flow. An initialization routine is required which specifies all possible reactions which may occur and gives the associated parameters for each reaction, such as activation energy. These parameters are detailed below. In the collision step itself, once a pair of particles has been chosen to collide (based only on the relative velocity), the algorithm checks whether a reaction may occur between the colliding pair. If a reaction is possible, with a certain probability, based again on the relative velocity, the pair reacts to produce other species such that momentum and energy are conserved. Dissociation and recombination reactions are possible, so that the total number of particles is not necessarily conserved. For simulations involving internal degrees of freedom such as rotational and vibrational energies, such effects may also be included in determining the reaction energy of the colliding pair.

The key question is with what probability does a given reaction occur. This is provided by the use of steric factors, or reaction probabilities. Ideally, these probabilities would be obtained as functions of energy levels and impact parameters experimentally, or analytically from quantum calculations. However, such data are not available for most reactions of interest. The practical alternative is to assume a computationally convenient form of the steric factor which reproduces the standard reaction rate equations for continuum, near-equilibrium flows. The constants in the steric factor model may then be obtained from the more readily available data for continuum reaction rates.

Following Bird's development [12], the steric factor is obtained by comparing the continuum reaction rate equation for the change in the number density of species A due to reactions with species B

$$\frac{dn_A}{dt} = -aT^b \exp\left(\frac{-E_a}{k_B T}\right) n_A n_B \tag{3.57}$$

with the collision theory rate equation

$$\frac{dn_A}{dt} = -n_A n_B (\sigma V_r)_{AB} \int_{\frac{E_a}{k_B T}}^{\infty} P_r F(e) de. \tag{3.58}$$

Here n_A is the number density of A, a and b are constants, E_a is the activation energy of the reaction, $(\sigma V_r)_{AB}$ is the mean total collision cross section of an equilibrium gas, P_r is the steric factor, and $F(\epsilon)$ is the distribution function for an equilibrium gas of the quantity $\epsilon = E_c/k_B T$, where

$E_c = \frac{1}{2}m_r V_r^2$ and m_r is the reduced mass of species A and B. For a gas with translational energy only, this distribution function is

$$f(\epsilon) = \frac{1}{\Gamma(\frac{5}{2} - \omega)}\epsilon^{3/2-\omega}\exp(-\epsilon) \tag{3.59}$$

where ω is the Variable Hard Sphere temperature exponent. If P_r is assumed to be of the form

$$P_r = \kappa(E_c/k_B)^{\alpha+\omega-3/2}(1 - E_a/E_c)^\alpha \tag{3.60}$$

for some constants κ and α, then the integral in the collision theory rate equation can be evaluated analytically, and by comparison with the continuum rate equation, these constants may be evaluated as

$$\alpha = b + \frac{1}{2} \tag{3.61}$$

and

$$\kappa = \frac{\sqrt{\pi}a\left(\frac{m_r}{2k_B}\right)^{1-\omega}}{2\Gamma(b+3/2)\sigma_{AB}}. \tag{3.62}$$

Here σ_{AB} is part of the collision cross section of molecules A and B which does not depend on the relative velocity. Thus if the continuum rate equation constants a, b and E_a are known, then the reaction probability P_r is known as a function of E_c. Note that for $E_c < E_a$, P_r is taken to be zero.

A total steric factor P_T for molecules A and B at a given translational energy E_c is found by summing the contributions P_i from all possible reactions between A and B. A reaction occurs with probability P_T, and it is reaction i with probability P_i/P_T. Once reaction i is chosen, the energy E_c of the reacting pair is increased by the heat of reaction. New velocities of the colliding particles are assigned in the usual way such that the momentum and updated energy are conserved.

3.4 Simplified Flows

Some of the limiting case flows described at the beginning of this chapter may be computed using a simplified version of the DSMC algorithm. These modifcations are now described.

3.4.1 Test Particle Method

This approach is used when a certain species or type of particle is present in a dominant background gas in such a small quantity that each particle may be considered independent and uninfluenced by the presence of any other particles of that type. Moreover, the effect of these special particles on the background gas may also be neglected. For example, titanium atoms in a sputtering reactor operating at very low pressures (described in detail in the next chapter). Here the titanium atoms travel from a source to a target through a background of argon gas. If the density of argon is low enough, there will be so few collisions of the high energy titanium with the argon atoms that the background gas conditions will not be noticeably effected. However, the effect of each collision on the titanium atoms is significant in determining the angular distribution of their velocities. The relatively small number of titanium atoms present in the gas at any given time makes it unnecessary to consider titanium-titanium collisions.

In this scenario it is unnecessary to compute the background gas as a collection of particles whose distribution function satisfies the Boltzmann equation. Instead it may be assumed that the distribution function is a Maxwellian at a constant density, temperature and stream velocity, and these three parameters describe the background completely. The problem is then that of solving the linear Boltzmann equation

$$\frac{\partial f}{\partial t} + v \cdot \nabla_x f = Q(M, f) \,, \tag{3.63}$$

where M is the background Maxwellian distribution. Here f is the velocity distribution function of the test particle species.

As with the full Boltzmann equation, if the time discretization is chosen small enough, the convection and collision steps may be decoupled. A similar argument to what is given above can be used to show that the probability of a collision between a test particle molecule and a background molecule is

$$P = n_{BG} < \sigma(V_r) V_r > \Delta t \,. \tag{3.64}$$

Here n_{BG} is the number density of the background gas, and

$$< \sigma(V_r) V_r > = \int_{\mathbf{R}^3} \sigma(V_r) V_r M(w) dw \tag{3.65}$$

where $V_r = |v - w|$ and v is the velocity of the test particle. The VHS model may be used for $\sigma(V_r)$.

The cell size may be chosen based on n_{BG} and $\sigma(V_{max})$, where V_{max} may be taken to be the largest velocity of a test particle likely to be encountered in the simulation. For a typical application of high energy particles leaving a source, this value is related to the energy of the source. These values may also be used to set the time step Δt so that P_{max}, the probability of a collision between the fastest test particle and the background is some fraction of unity.

The simulation is run just as DSMC, with the exception that each test particle in a cell collides with probability $P(v)$ depending only on its velocity. If a collision is accepted, then the collision partner is sampled from M. Only the post collision velocity of the test particle is computed.

The use of a cell structure is not vital, but it is convenient to retain so that the test particle method may be included as a collision option in a full DSMC code. Moreover, the cell structure allows that the background gas have a spatially varying Maxwellian distribution. A variation of this procedure which does not use the cell structure is described in Chapter 8.

3.4.2 Free Molecular Flow

Free molecular flow was discussed at the end of Chapter 2. Except for the special case of diffuse reflection at the walls, for which a view factor method may be suitable, Monte Carlo ray tracing methods are generally used. As explained in Section 2.5, free molecular flow refers to the limiting case when the gas is so rarefied that particle-particle collisions no longer play a significant role in determining the gas dynamics. In this case the transport equation becomes simply

$$\frac{\partial f}{\partial t} + v \cdot \nabla_x f = 0\,. \tag{3.66}$$

This equation may be solved with the DSMC algorithm by turning off the collision step. The cell size and time step restrictions must be based on other criteria than mean free path (which is now infinite). These should be chosen to be some fraction of the macroscopic length scales of the domain and the mean residence time in the domain.

Chapter 4

Gas Transport Simulations

A brief description of the most important microelectronics manufacturing equipment for gas transport was presented at the beginning of Chapter 2. Figure 2.1 shows the regimes within which the various reactors operate. The flow in the reactors in the hydrodynamic regime is suitably modeled by the Navier-Stokes equations with no-slip boundary conditions. As the Knudsen number increases, however, rarefied gas effects begin to appear. For reactors with regions that operate in the near hydrodynamic regime, such as the LPCVD and UHV-CVD reactors, the fluid dynamic equations must be supplemented with slip boundary conditions. Reactors operating in the transition regime, such as the UHV-CVD and sputtering reactors, must be modeled by the Boltzmann equation. In addition, in all wafer processing applications, Knudsen regimes involving free molecular flow may arise if small scale features of the microelectronic device structure, such as trenches, are modeled.

We begin this chapter with a numerical study of the slip boundary conditions necessary for calculations in the near hydrodynamic regime. The key issue here is to determine the slip coefficients. While some theory exists which predicts the coefficients in an idealized setting, more reliable results are obtained from experiment. Thus we present the results of a series of numerical experiments whereby the Boltzmann equation is solved using the DSMC method and the slip coefficients are computed from the solution.

Next we turn to the simulation of the UHV-CVD and sputtering reactors using DSMC. The gas flow in the UHV-CVD reactor lies just on the border between the near hydrodynamic and the transitions regimes. The results of the simulation show significant deviations from what continuum theory would predict. These effects include enhanced species separation as well as a larger pressure drop.

The flow in the sputtering reactor lies squarely in the transition regime. We present a detailed study of this reactor which determines, among other results, the effect of the sputtering atoms on the background gas as a function of pressure. This in turn may be used to decide when the simplified Test Particle method (see Chapter 3) may be used in place of the full DSMC simulation. An important result of the sputtering reactor simulation is the determination of the angle distribution of sputtering particles as they arrive at the target surface. This distribution is necessary as a boundary condition for a feature scale free molecular flow simulation of trench filling.

4.1   Near Hydrodynamic Effects

4.1.1   Determination of Slip Coefficients

The goal here is to determine the various slip coefficients which appear in the velocity and temperature slip boundary conditions in Equations (2.38) and (2.39):

$$\vec{t} \cdot \vec{v}\big|_{z=0} = \zeta^{vv}\vec{n} \cdot \nabla(\vec{t} \cdot \vec{v}) + \zeta^{vT}\vec{t} \cdot \frac{\nabla T}{T} + \sum_{i=1}^{N} \zeta_i^{vp}\underline{t} \cdot \vec{X}_3^i \qquad (4.1)$$

$$T_{wall} - T\big|_{z=0} = \zeta^{TT}\vec{n} \cdot \frac{\nabla T}{T} + \sum_{i=1}^{N} \zeta_i^{Tp}\vec{n} \cdot \vec{X}_3^i \qquad (4.2)$$

This is achieved by expressing each coefficient as a simple function of the main physical properties of the flow (the mean free path λ and the temperature T) multiplied by a dimensionless slip coefficient. It is generally assumed that the slip is proportional to the mean free path, so that as the gas becomes more dense and the hydrodynamic regime is approached, the slip effects become negligible.

The dimensionless slip coefficients are evaluated through a series of computational experiments. Flows which highlight the various slip effects are simulated with DSMC, and the resulting slip is determined. The computed slip is then matched with the predicted slip from Equations (4.1) and (4.2) in order to determine the dimensionless slip coefficient. The experiment may be repeated at different flow conditions to determine the dependence of the dimensionless slip coefficient on pressure, species concentration, etc.

The experiments consisted of computing either plane Poiseuille flow or plane Couette flow. The Poiseuille flow is generated by two large reservoirs maintained at different states connected by a channel consisting of an upper and lower plate. This is illustrated in Figure 4.1 (a). The length of the channel is finite but long enough that edge effects at the inlet and outlet can

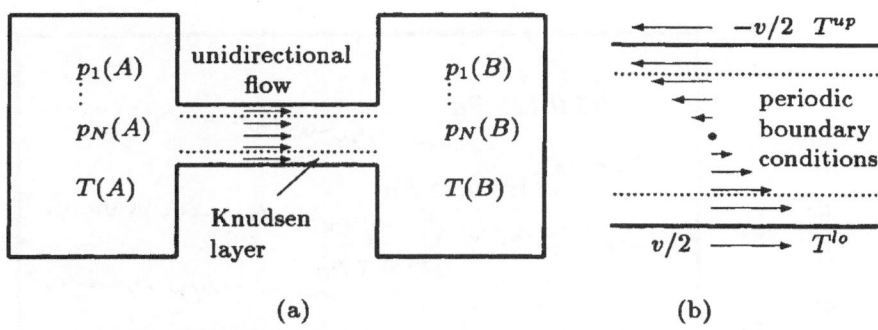

Figure 4.1: (a) Poiseuille flow geometry, (b) Couette flow geometry

be neglected. The reservoirs are filled with a gas mixture of species $1, ..., N$ at partial pressures $p_1, ..., p_N$ and temperature T. All relative differences of the variables of state between A and B should be small compared to one. In the near hydrodynamic regime, characterized by $0.01 < Kn < 0.1$, the layers between the bulk flow and the surface, where the deviations from equilibrium are very large, are well separated. These Knudsen layers have an extension of about one mfp. The Poiseuille flow geometry is used to compute the thermal creep and diffusive creep coefficients.

Couette flow is generated between two parallel plates such that the flow gradients exist only in the normal direction to the plates. Because the flow is constant in the tangent direction, periodic boundary conditions may be used. The geometry for this flow is illustrated in Figure 4.1 (b). This setting allows the computation of the velocity slip and temperature jump coefficients. In the case of velocity slip, the two plates are in relative motion, while for temperature jump the plates are stationary, but a temperature difference is imposed.

The slip information obtained from the computation of flow between flat plates which have well separated Knudsen layers is also a good approximation to a flow along an arbitrary surface where the radius of curvature and the distance to other surfaces are large compared to the mean free path.

Thermal Creep Coefficient

Following the parameterization found in [92], we formulate the thermal creep coefficient ζ^{vT} in terms of the dimensionless coefficient $\widehat{\zeta}^{vT}$, the mean free

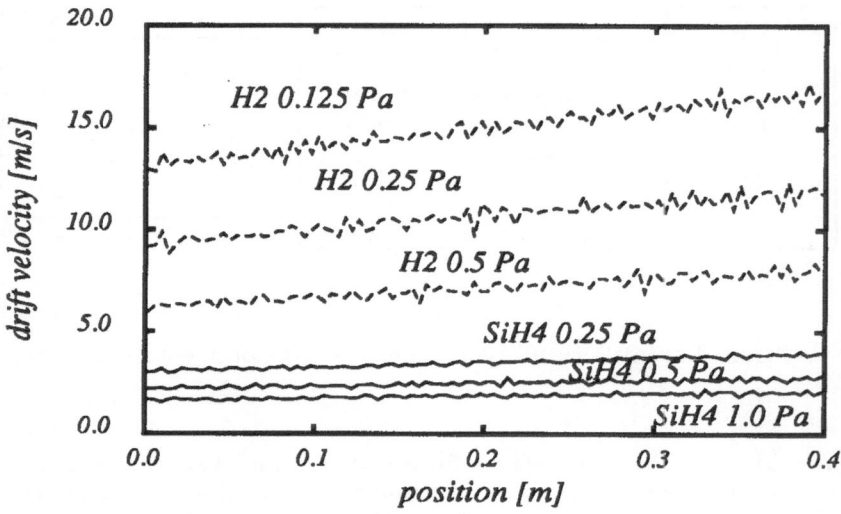

Figure 4.2: Drift velocity from thermal creep effect. The drift is from the colder to the warmer end.

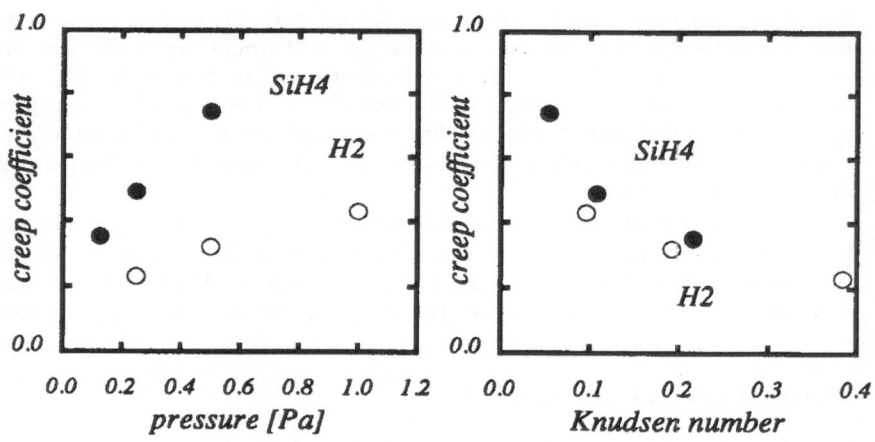

Figure 4.3: Dimensionless thermal creep coefficient

path λ, the temperature T and the thermal velocity

$$v_{th} = \sqrt{\frac{8k_B T}{\pi m}} \ . \tag{4.3}$$

The thermal accommodation coefficient α of the wall must also be taken into account. Chapman-Engskog theory indicates that the thermal and diffusive creep coefficients are proportional to α. The thermal creep coefficient is taken to be of the form

$$\zeta^{vT} = \lambda v_{th} \ \alpha \ \widehat{\zeta}^{vT} \ . \tag{4.4}$$

A value of $\widehat{\zeta}^{vT} = 0.4$ is suggested in [92].

The Poiseuille flow geometry shown in Figure 4.1 (a) was used in the DSMC simulation to determine the dimensionless thermal creep coefficient. The 2-D channel was taken to have length 0.4 m and height 0.1 m. The reservoirs were represented as Maxwellian distributions at each end of the channel with temperatures $300°K$ and $400°K$ respectively. The density of each distribution was chosen so that the pressure would be constant throughout the channel. The walls were fully accommodating ($\alpha = 1$), and a linear temperature profile from $300°K$ to $400°K$ was specified along the walls connecting the reservoirs. Two separate sets of experiments were conducted, one in which the channel was filled with hydrogen gas H_2, and the other in which silane SiH_4 was used. In both cases the appropriate variable hard sphere model was used. After a transient period, a steady flow evolved from the colder to the warmer end.

The velocity in the channel is constant along the cross section (notably different from the parabolic velocity profile which arises in pressure driven flows) and increases slightly along the channel with the temperature. The computed velocities at different pressures are shown in Fig. 4.2. For this problem, Equation (4.1) reduces to

$$v = \zeta^{vT} \frac{1}{T} \frac{\partial T}{\partial x} \tag{4.5}$$

where $\partial / \partial x$ is the derivative along the channel. Together with Equation (4.4) this may be used with the computed velocity data to determine the dimensionless slip coefficient. The results are shown in Figure 4.3. This shows a pressure as well as a species dependence. Also a slight temperature dependence is visible. The value $\widehat{\zeta}^{vT} = 0.4$ is a rough estimate, but clearly a model is needed to account for the dependence on Knudsen number, molecular species and temperature.

Diffusive Creep Coefficient

For the diffusive creep coefficient ζ_i^{vp} we propose the parameterization

$$\zeta_i^{vp} = \lambda_i \ \frac{m_i v_{th,i}}{k_B T} \ \alpha \ \widehat{\zeta}_i^{vp} \ . \tag{4.6}$$

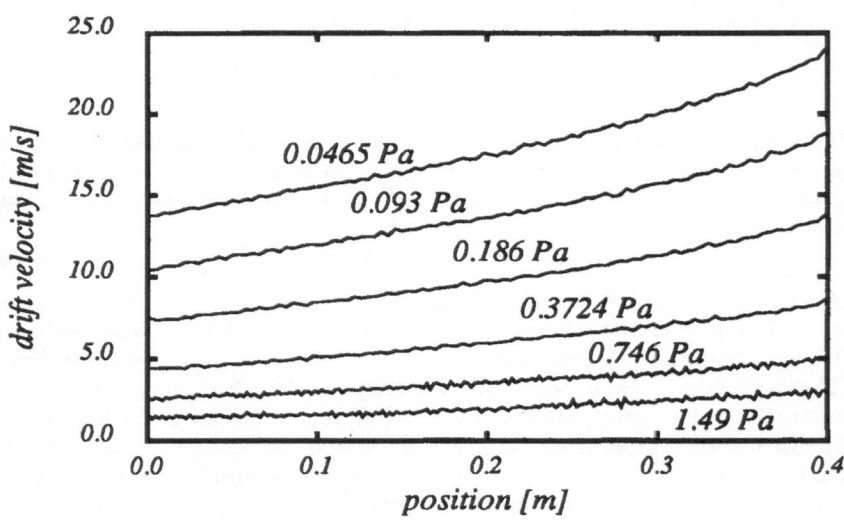

Figure 4.4: Drift velocity from diffusive creep effect. The drift is from the side enriched with heavier species to the side enriched with the lighter species.

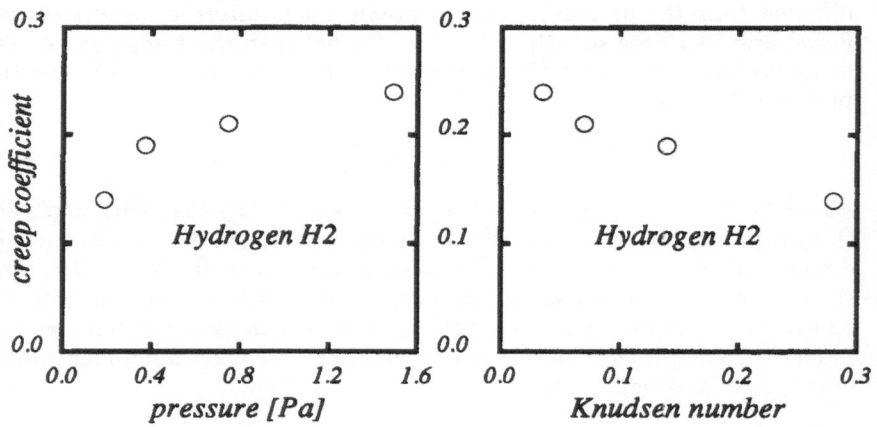

Figure 4.5: Dimensionless diffusive creep coefficient for hydrogen.

This is somewhat different that what is found in [92] .

An experimental set up similar to the thermal creep case was used to determine the diffusive creep. In this case, however, the wall and reservoir temperatures were held constant at $300°K$. The dimensions of the channel were the same. The key element of this simulation was the concentration gradient. This was achieved by using a gas mixture of hydrogen and silane. At the left entrance of the channel the reservoir Maxwellian consisted of 0.7 mole fraction SiH_4 and 0.3 mole fraction H_2, while the reservoir at the right end was 0.3 mole fraction SiH_4 and 0.7 mole fraction H_2. Again the walls were diffusively reflecting. After a transient period, a steady flow evolved from the side enriched with the heavier species to the side enriched with the lighter species.

The calculated velocities at different pressures are shown in Figure 4.4. For this problem, Equation (4.1) reduces to

$$v = \zeta_1^{vp} \frac{1}{\rho_1} \frac{\partial p_1}{\partial x} + \zeta_2^{vp} \frac{1}{\rho_2} \frac{\partial p_2}{\partial x} . \tag{4.7}$$

Together with Equation (4.6) this may be used with the computed velocity data to determine the dimensionless slip coefficients. The slip contribution from hydrogen, species 1, is an order of magnitude larger than that of silane. Thus only the first term of the above sum was used. The computed velocity has a significant dependence on x which was difficult match with the spatial dependence of the slip model. A rough fit was possible, and the resulting slip coefficients are shown in Figure 4.5.

Velocity Slip Coefficient

The standard parameterization of the velocity slip coefficient ζ^{vv} is

$$\zeta^{vv} = \lambda \frac{2 - \alpha}{\alpha} \widehat{\zeta}^{vv} . \tag{4.8}$$

Considerably more theoretical work exists in the area of the velocity slip coefficient (establishing the above form and evaluating the dimensionless coefficient for various collision models) than for the creep coefficients. A summary of analytical techniques which have been used can be found in [20]. The results show that $\widehat{\zeta}^{vv}$ is usually around one.

To show the effect of slip in a numerical experiment, the Couette flow geometry of Figure 4.1 (b) with periodic boundary conditions and full accommodation $\alpha = 1$ was used. The velocity profiles are shown in Figure 4.6. The fact that they are not linear is due to the changing viscosity across the channel which results from the temperature rise of the compressible gas. The velocity slip coefficient is evaluated as follows. First the Navier-Stokes equations

$$(\mu \, u_y)_y \;\; = \;\; 0 \tag{4.9}$$

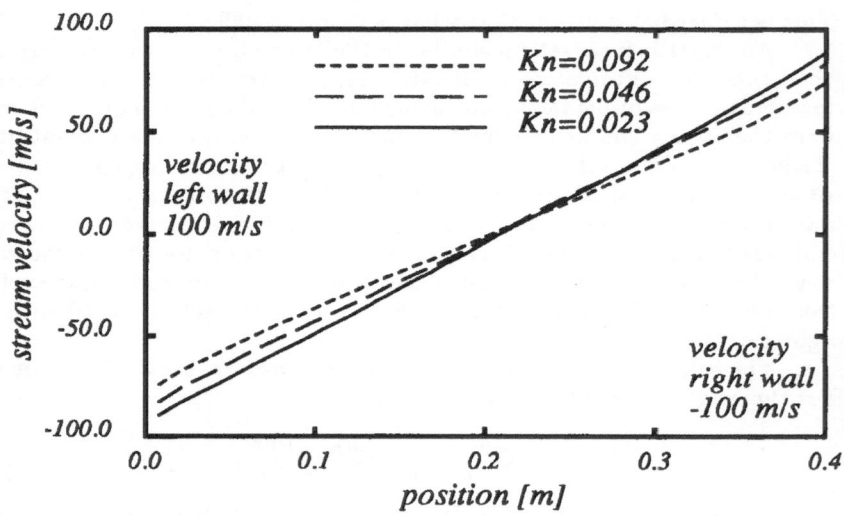

Figure 4.6: Velocity profiles in the Couette geometry with slip effect. The gas is hydrogen.

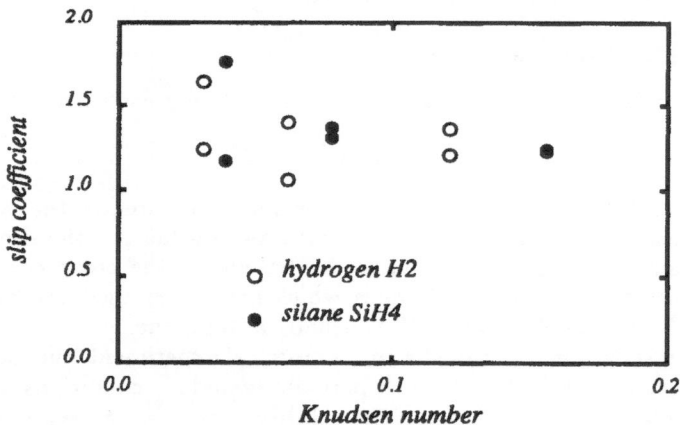

Figure 4.7: Dimensionless velocity slip coefficient.

$$(\mu\, u\, u_y + \kappa k_B/m T_y)_y \;=\; 0$$

with slip boundary conditions

$$u(y = 0) \;=\; \zeta^{vv}|_{y=0}\,\left.\frac{\partial u}{\partial y}\right|_{y=0} \qquad (4.10)$$

$$u(y = L) \;=\; -\zeta^{vv}|_{y=L}\,\left.\frac{\partial u}{\partial y}\right|_{y=L}$$

$$T(y = 0) \;=\; T^{wall}$$

$$T(y = L) \;=\; T^{wall}$$

are solved. Here it is assumed that the temperature gradient at the wall is small enough that the temperature slip effects may be ignored. Then the solution is fit to the numerical velocity profile from the DSMC calculation to determine the dimensionless velocity slip coefficient. The viscosity and the thermal conductivity are given by the VHS formulas from Equations (2.53) and (2.55).

The Temperature Jump Coefficient

The temperature jump coefficient can be similarly parameterized as

$$\zeta^{TT} = \lambda\,\frac{2-a}{a}\,\hat{\zeta}^{TT} \qquad (4.11)$$

As with the velocity slip, some theoretical work has been carried out [20]. These results predict the dimensionless temperature jump coefficient to have a value around 2.

Again the dimensionless coefficient can be obtained from a numerical experiment with DSMC. For this purpose, the Couette flow geometry of Figure 4.1 (b) with zero wall velocity but different temperatures of the plates of $300°K$ and $1000°K$ respectively was used. Two sets of experiments were conducted: one in which the domain was filled with the very light H_2, and one in which the very heavy WF_6 was used. As in the case of the velocity profile used in the determination of the velocity slip, the temperature profiles are not linear, as can be seen in Figure 4.8. This comes here from the temperature dependence of the thermal conductivity. The evaluation of the numerical experiment for the slip can be done as follows. First the energy transport equation of the Navier-Stokes system is solved

$$(\kappa k_B/m T_y)_y = 0 \qquad (4.12)$$

with boundary condition

$$T(y = 0) \;=\; \zeta^{TT}|_{y=0}\,\left.\frac{\partial T}{\partial y}\right|_{y=0} \qquad (4.13)$$

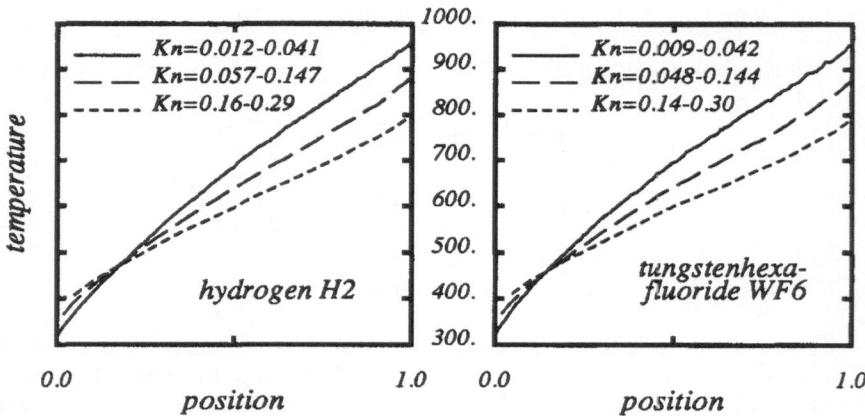

Figure 4.8: Temperature profile between two plates at 300° and 1000° respectively.

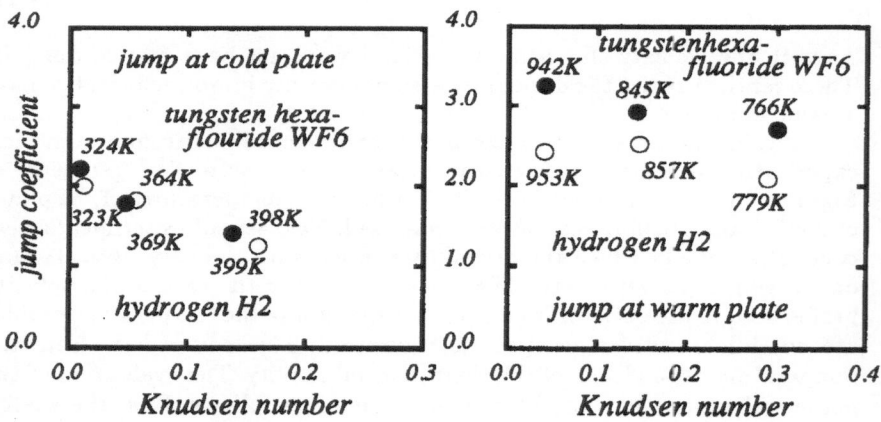

Figure 4.9: Values for the dimensionless temperature jump coefficient

$$T(y = L) = \zeta^{TT}|_{y=L} \left.\frac{\partial T}{\partial y}\right|_{y=L} .$$

The solution is then fit to the DSMC results to determine the value of $\widehat{\zeta}^{TT}$.

The results of the numerical experiments are shown in Figure 4.9. Since there is a jump at the warmer and the colder wall, there are results for two different temperature levels around $300° K$ and $1000° K$. The results show the significant temperature and species dependence as well as the dependence on the Knudsen number.

Conclusions

The slip effects described above are all first order in the Knudsen number. Information on the order of a transport phenomenon comes only from analyzing the Boltzmann equation. The Onsager theory which was used to suggest possible slip effects does not indicate their size. It is assumed that the remaining slip coefficients, such as the temperature slip due to divergence of the flow (Equation 2.39), are of lesser important. Thus they were not investigated.

The numerical experiments described above provide values for the dimensionless slip coefficients. Furthermore, they show the dependence of these coefficients on molecular species, temperature and Knudsen number. These dependencies are so strong, that they should be contained in a model. Such models, however, have yet to be firmly established. It may be possible to develop suitable models with a phenomenological approach using data from numerical experiments or from further analysis of the Chapman-Engskog solution of the Boltzmann equation.

The uncertainty in the values of the slip coefficients makes it difficult to use slip boundary conditions with a hydrodynamic fluid equation solver. Estimates for the coefficients exist, but the error is still large.

4.1.2   The Diffusion Coefficient

The slip effects discussed above are of first order in Knudsen number. As the Knudsen number increases, some second order effects may appear, such as those described by the Burnett equations. They lead to a modification of the currents [54]. An example of this is the diffusion current $\vec{j}_1$ of a binary mixture. With increasing Knudsen number the flux corrections arise. In the limit of the Knudsen regime, however, where there is no particle interaction, the diffusion can be described by the Knudsen diffusion coefficient (see Section 2.5).

If one fits the uncorrected expression of the ordinary diffusion current to the diffusive transport, one obtains a kind of transition regime diffusion

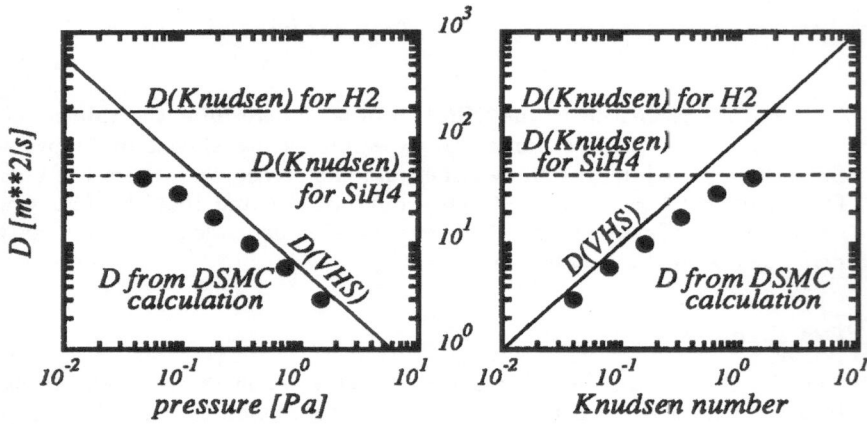

Figure 4.10: Diffusion coefficient in rectangular channel

coefficient [24], [23]. Although this method is not rigorous, it has the advantage that it shows, in form of the transition regime diffusion coefficient, when a deviation in the transport behavior compared to the hydrodynamical regime arises.

The computational experiment used to determine the diffusive creep coefficient can also be used to evaluated the transition regime diffusion coefficient. This is done by fitting the numerical results to the hydrodynamic diffusive current given by Equation (2.6) and the (uncorrected) expression (2.27).

$$\vec{j}_1 = \rho_1(\vec{v}_1 - \vec{v}) = \mathcal{D}_{12}\frac{n^2}{\rho}m_1 m_2 \nabla \left(\frac{n_2}{n}\right) \tag{4.14}$$

The species velocity $\vec{v}_1$ is easily obtained from the DSMC calculation. The resulting binary diffusion coefficients for the mixture of SiH_4 and H_2 are shown in Figure 4.10. This figure also shows the Knudsen diffusion coefficients which were calculated from Equation (2.73).

Finally, the figure shows the theoretical value for the binary diffusion coefficient calculated from the VHS model. This is computed in the following way. The mixture viscosity is known to be a relatively complicated function of the mole fraction of the species and the single species viscosities. The contribution of the lighter species is enhanced. A pragmatic ansatz in the frame of the VHS model is to substitute the average of the VHS diameters of the components of the mixture for the diameter in Equation (2.51), i.e. $0.5(d_{1,VHS}(V_r) + d_{2,VHS}(V_r))$ for a binary mixture. The result is some mixture viscosity which is, however, mole fraction independent. The diffusion coefficient of mixtures is known to be only slightly dependent on the mole fraction. This surprising fact makes it possible to give good model estimates for diffusion coefficients. Here it is obtained from the self diffusion coefficient given by Equation (2.55) by substituting in the reduced mass $m \rightarrow 2m_r$ and the value for the mixture viscosity.

In the hydrodynamic and near hydrodynamic regimes, the diffusion coefficient from the computational experiment is very close to the calculated continuum value. This corresponds to the second order nature of the flux correction. In the transition regime, however, the deviation of the transition regime diffusion coefficient from the continuum value becomes significant.

4.2   UHV-CVD Reactor

We now turn to the simulation of an Ultra High Vacuum Chemical Vapor Deposition (UHV-CVD) reactor. As discussed in Section 2.1, the flow regimes in this reactor may range from near hydrodynamic to transition to near Knudsen flow. The nonequilibrium effects which arise may have a significant effect on the determination of the flow. Such effects can only be modeled by solving the Boltzmann equation. Therefore the DSMC method is used for the simulation. This approach was first studied by Coronell et. al. in [23], [24].

The UHV-CVD reactor operates at significantly lower pressures than other CVD reactors. The reduction in pressure has a number of advantages. The flow is well in the laminar regime, diffusion dominates the species transport and leads to very good uniformity over the wafer, and the slip flow along the boundaries even reduces the effect of autodoping (evaporation of dopant into the gas phase).

However, a variety of transport effects arise which lead to an unexpected behavior. In the near hydrodynamic regime the effects of pressure and thermal diffusion appear in the continuum flow. At the boundaries the additional effects of velocity slip, thermal and diffusive creep, and temperature jump can be observed. In the transition flow there are a variety of unnamed effects related to the nonequilibrium character of the velocity distribution function.

To demonstrate these effects, a flow simulation was conducted in a model UHV-CVD reactor. Figure 4.11 shows the draft of the geometry of an axisymmetric reactor of length $0.4\ m$ and diameter $0.1\ m$. The domain was resolved into 40 cells in axial direction and 20 equivoluminal, annular cells in radial direction. The calculation was done for the three inlet pressures of $2.0\ Pa$, $0.6\ Pa$ and $0.2\ Pa$ and flowrates of $32.5\ sscm$, $32.6\ sscm$ and $14.9\ sscm$ respectively (the first two total flow rates are quite similar, but the partial flow rates of each species are quite different as a consequence of the different pressures). The computation time on a HP730-series work-

P_{inlet} [Pa]	P_{outlet} [Pa]	Q_{SiH_4} [sscm]	Q_{H_2} [sscm]	Kn	Δt [s]	steps
2.0	1.45	11.8	20.7	0.03-0.1	7.7×10^{-6}	6000
0.6	0.20	7.5	25.1	0.1-0.3	7.7×10^{-6}	6000
0.2	0.06	2.9	12.0	0.3-1.0	5.5×10^{-6}	4000

Table 4.1: Data of the UHV/CVD simulations. The units of the flowrate are standard cubic centimeters per minute [sscm].

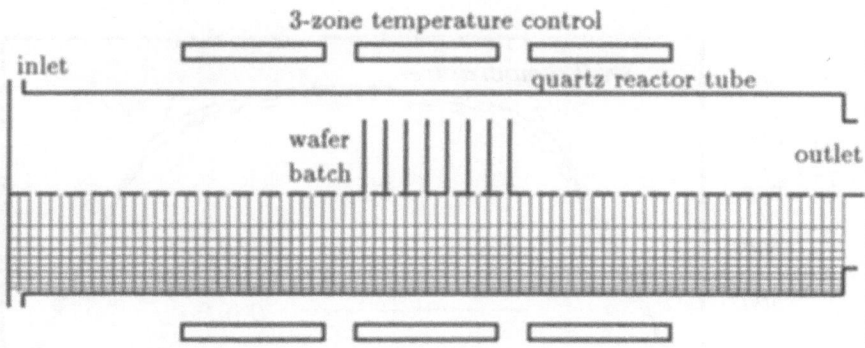

Figure 4.11: Geometry of the UHV-CVD equipment and computational grid. The grid has 80 cells in axial and 100 cells in radial direction.

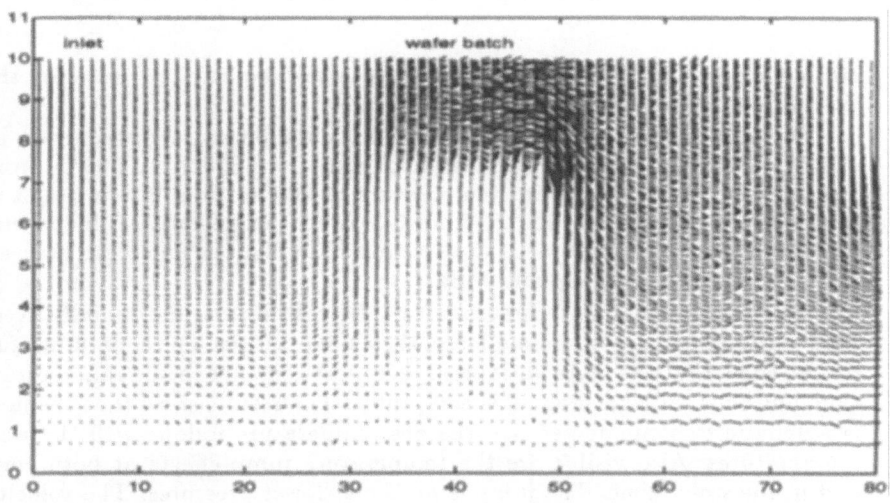

Figure 4.12: Velocity vectors in the reactor

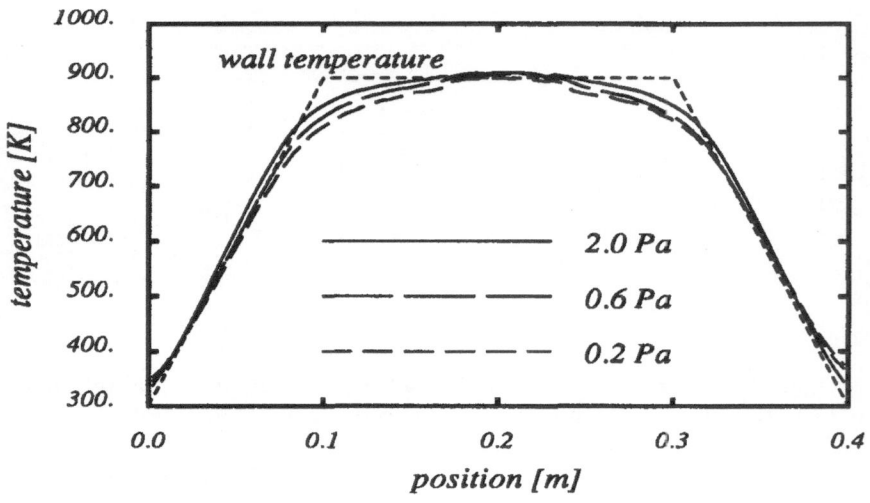

Figure 4.13: Temperature distribution along wall and in the gas

station ranged from less than an hour to a few hours. The data for the simulations are summarized in Table 4.1.

In all the simulations, the walls had a temperature distribution as shown in Figure 4.13, namely a linear increase from $300° K$ to $900° K$ from the inlet to 0.1 m, a constant temperature of $900° K$ from 0.1 m to 0.3 m and a linear decrease to $300° K$ from 0.3 m to the exhaust. Full thermal accommodation was assumed at the walls. Finally, the inlet gas was an equal molar mixture of H_2 and SiH_4.

Figure 4.13 shows the first result of the calculation, the radially averaged gas temperatures as a function of the axial position. The computational results show that the state variables are almost constant in the radial direction, and thus radial averaging is acceptable. The results show that the flow is adiabatic to some degree, i.e. the gas temperature follows well the wall temperatures. Also visible are the temperature jump effects at both ends and in the wafer zone. The jump is higher at lower pressures. The velocity slip effect can be seen in Figure 4.12 which shows the velocity vectors for a typical flow.

The most noticeable effect in this flow is the large pressure drop, which leads to an outlet pressure much smaller than the inlet pressure. Figure 4.14 shows a typical 2-D pressure distribution. The pressure drop limits the mass flow through the reactor. This was visible in the simulation at the lowest pressure, where the desired mass flow of around 14.9 $sscm$ for a

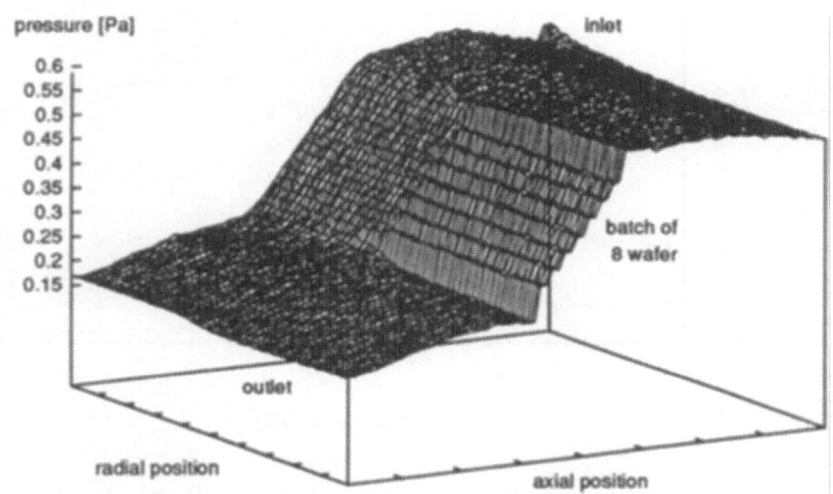

Figure 4.14: Typical 2-D pressure distribution in the reactor

fixed inlet pressure could not be achieved. The pressure drop leads further to the problems addressed in Section 2.4 concerning setting the inlet and outlet boundary conditions. For the simulation, a flux weighted Maxwellian distribution given by Equation (1.63) was used. Based on the fixed inlet pressure and fixed flow rate, a drift velocity for the inlet was calculated from mass conservation. The relatively small drift velocity had only minor influence on the problem. For the outlet, pressure, temperature and drift velocity were initially unknown. In the initial simulations, the pressure jump from the last cell of the domain to the outlet state was monitored. The main adjustments were done in the particle density of the outlet state to achieve the desired flowrate. Two or three iterations were necessary to provide a satisfactory result in terms of the value of the flow rate and the continuity of the pressure from the last cell of the domain to the outlet state. The smallness of the inlet cross section, the isothermal character of the flow and the small drift velocity greatly simplified the iteration procedure.

Figure 4.15 shows the radially averaged pressure for the three simulations including the pressures of the inlet and outlet states. The first surprising effect is the pressure increase in the flow direction. This increase is correlated with the temperature increase and shows the nonequilibrium character of the flow. In the language of slip effects, this is the thermal creep effect which is directed opposite to the flow from the inlet to the distance

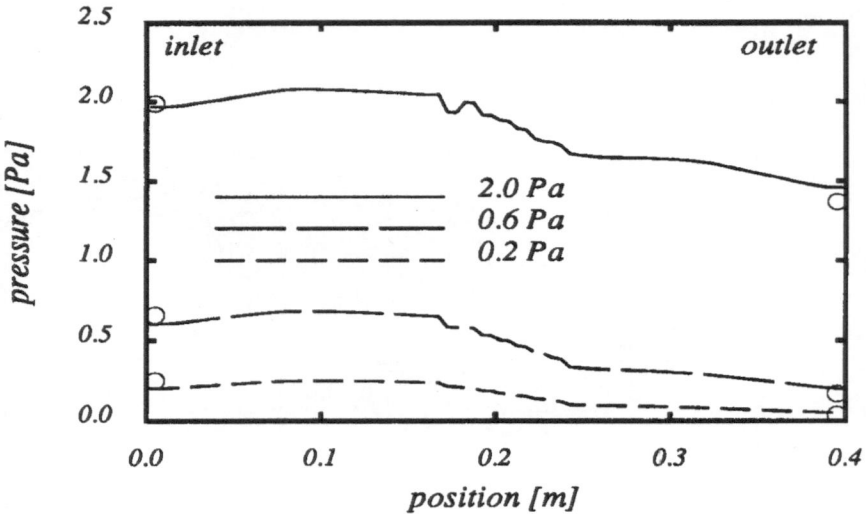

Figure 4.15: Radially averaged pressure distribution in the reactor. Remarkable is the pressure increase in flow direction due to the thermal creep. The circles indicate the values of the pressure chosen for the drifted Maxwellian at inlet and outlet.

$0.1\,m$. The analysis of the thermal creep effect in Section 2.1 provided examples for the size of this effect for a temperature gradient much smaller than in this example.

After the initial pressure increase, the pressure decreases slightly in the flow direction until the position of the first wafer. The following pressure drop is very strong due to the much smaller cross section for the flow and the correspondingly larger resistance. Behind the wafer region there is again a slight pressure drop. This drop increases from the point at $0.3m$ to the outlet in the region where there is a negative temperature gradient. This is again a thermal creep effect, but it now drives the molecules in flow direction. Figure 4.16 shows the radially average of the axial component of the drift velocity.

Finally we consider the individual species properties of the flow. There are strong mass separation effects in the flow of the SiH_4 and H_2 mixture due to their quite disparate masses. This can be seen in Figure 4.18. Also the flow conductance is very different for the two species. These effects depend on the physical aspects of the domain as well as on the choice of the boundary states with their physical implications. In principle the presence of multiple species creates yet another factor which must be included in

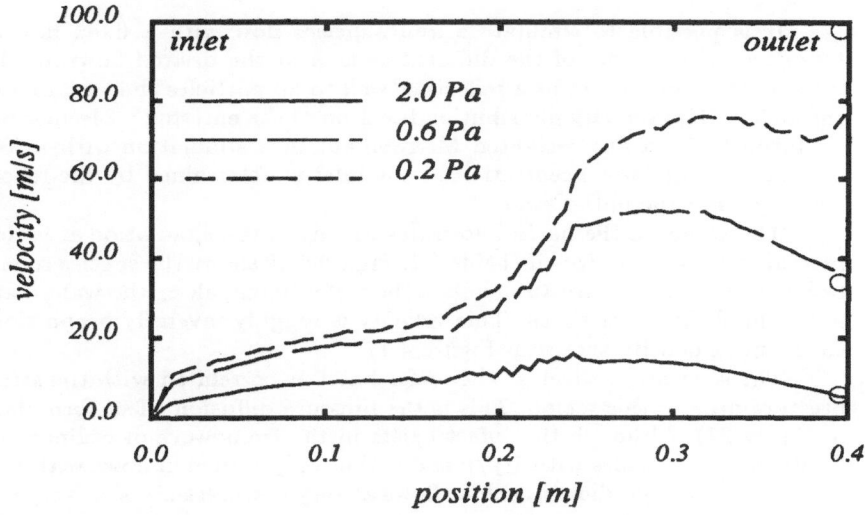

Figure 4.16: Radially averaged axial component of drift velocity along the reactor.

the boundary condition iteration process. In the present example, however, the species mole fractions set at the outlet had a negligible effect on the other flow properties. This made it possible to first iterate on pressure and total flow rate, then separately iterate the outlet concentrations to achieve continuity.

Specifying the inlet concentrations also requires some consideration. There exist two possibilities for defining the inlet conditions for a multiple species flow. The first is to fix the mole fraction but not the partial flowrate, while the second fixes the partial flowrate but not the mole fraction.

The simulation of the inlet as molecules entering from a flux weighted Maxwellian models the first case. Each species should be assigned its own Maxwellian parameterized by its individual drift velocity. The number of particles of each species entering the domain per unit time is set to match the desired mole fractions. The drift velocities may then be iterated to produce continuous results. In a simulation with specified molar concentration, the flowrate is determined by the reactor geometry and the outlet state.

The results in Figure 4.18 show a slight deviation from the desired 0.5 mole fraction near the inlet. This was the results of using the same drift velocity (corresponding to a single flux weighted Maxwellian) for each species. This simplifies the iteration procedure somewhat at the cost of a relatively small error in mole fraction.

It is possible to simulate a multi-species flow with a fixed flowrate by injecting molecules of the different species at the desired flowrate. The inlet must, however, act as a reflecting wall to all particles that try to leave through it. The velocity distribution function of the entering molecules may be taken to be a flux weighted Maxwellian. In a simulation with a fixed flowrate, the molar concentration at the inlet is determined by the reactor geometry and the outlet state.

The values of the partial flowrates arising in the simulation at various pressures are summarized in Table 4.1. Figure 4.18 shows the species separation effects in the reactor tube before the wafer batch, along the wafer batch and behind the wafer batch. This velocity is roughly inversely proportional to the mass density shown in Figure 4.17.

The separation effect at the wafer batch is correlated with the strong pressure drop in this region. This is the pressure diffusion effect formulated in Eq. (2.27). Although this effect exists in the framework of ordinary hydrodynamics, it scales with $\nabla p/p$ and is thus only visible in flows with large relative pressure gradients, such as flows at very low pressure or at very high stream velocity. Diffusive creep may also play a role in the mass separation due to the large concentration gradient.

The separation effects in front and in back of the wafer batch are of a different nature. They show a different behavior at high, medium and low pressure, but they are also correlated with the temperature gradient in the gas. This temperature effect leads to thermal diffusion, which drives the heavier species in the colder region. Since this is a bulk effect, it will increase with increasing pressure. At the end of the reactor, the mass separation at the high pressure indicates clearly the presence of thermal diffusion: the concentration of the heavier species increases towards the end. At the lowest pressure, however, the pressure diffusion overwhelms the effect of the thermal diffusion, and the concentration of the heavier species decreases. The same effect, but in the opposite direction, is visible at the inlet.

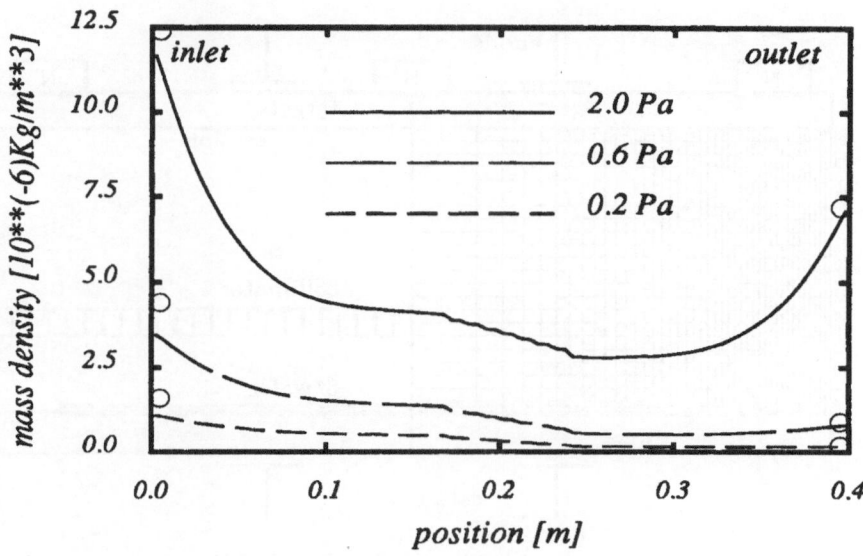

Figure 4.17: Radially averaged mass density along the reactor. The circles indicate the values of the mass density chosen for the drifted Maxwellian at inlet and outlet.

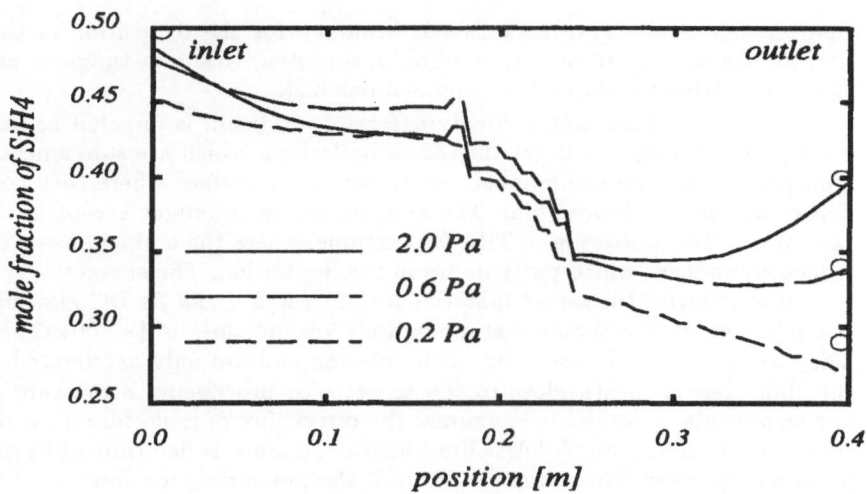

Figure 4.18: Radially averaged molar concentration along the reactor

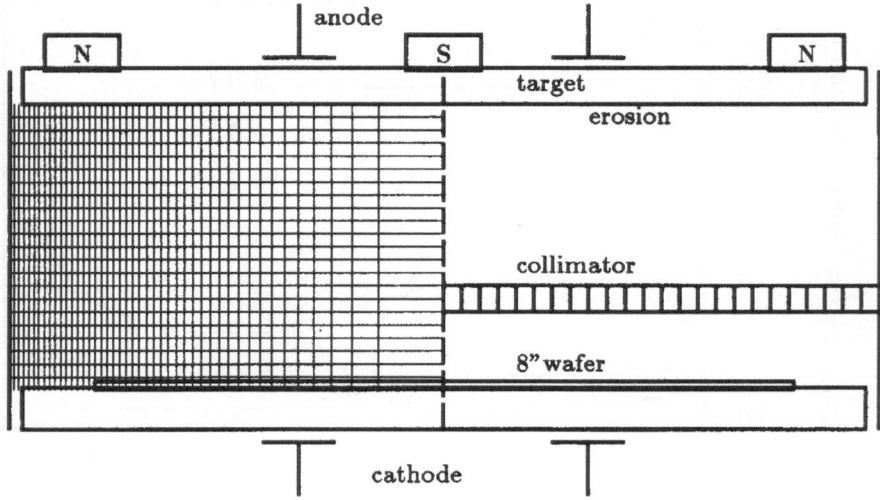

Figure 4.19: Geometry and computational grid of the sputter reactor.

4.3   Sputtering Reactors

Sputter deposition is an established technique for the deposition of thin film metals such as titanium, aluminum, tungsten, titanium-tungsten and titanium nitride for chip wiring applications [99].

In sputter deposition a highly energetic ion beam is directed against a target to dislodge (sputter) the target molecules, which are subsequently transported through the ambient gas to the wafer surface where they condense to form the desired film. The ambient gas is argon gas except in the case of reactive sputtering of TiN, for example, where the ambient has a nitrogen component which participates in the deposition. The energetic Ar^+-ions used to strike the target material are generated by an Ar DC-glow discharge in the reactor chamber at moderately low pressures of $1-100\ mTorr$. They are generated in the bulk of the plasma and are only accelerated in the thin plasma sheath close to the target. For processing, a pressure as low as possible is desired to maximize the direct flux of molecules from the target to the substrate. A lower limit for the pressure is determined by the plasma properties. The plasma vanishes if the pressure is too low.

Currently most sputtering is magnetron sputtering. Figure 4.19 shows a sketch of a magnetron sputter reactor. In this technique a magnetic field near the target surface is created which confines the electron trajectories to

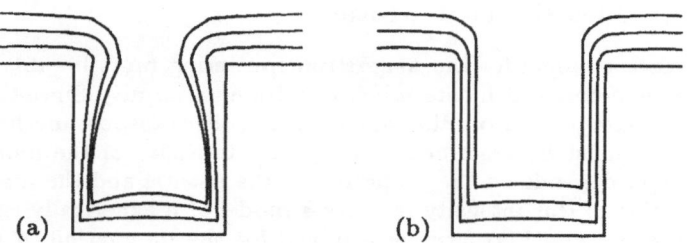

Figure 4.20: (a) shows the filling of a trench with sputter deposition and void formation. This is the consequence of an undirected angular distribution. (b) shows the filling of the trench without void formation which can be achieved with a forward directed angular distribution.

cyclotron orbits. This allows the pressure to be reduced to about 1 $mTorr$ while still maintaining the same level of ionization. The proper design of the magnetic field furthermore allows local maxima in the ionization density to be created such that the resulting source of sputtered molecules at the target surface has a spatial distribution which leads to greater uniformity of deposition on the wafer.

Some wiring integration schemes require high aspect ratio (depth divided by width) micro features such as vias or contact holes to be filled. In modern processes with extremely high aspect ratios, conventional sputtering processes cannot provide sufficiently large step coverage (layer thickness half way down the wall inside the feature divided by layer thickness at the top of wafer). This is because the angular distribution of the sputtered molecules is too wide. This leads first to an increased deposition at the entrance of the feature, then to a bottleneck, and finally to void formation. Figure 4.20 shows the profile evolution in a trench structure from molecules with unrestricted and restricted angular distributions.

The introduction of the collimated sputter technique has brought the general advantages of sputtering to modern chip generations of 64M and 256M technologies which have feature sizes on the order of $0.25\mu m$ and high aspect ratios. In this technique, a plate with a collection of holes positioned between target and substrate plane acts as a filter for the angular distribution of the sputtered particles as can be seen in Figure 4.19. The selectivity of the filter depends on the aspect ratio of the holes in the collimator plate. The price for the superior step coverage is the reduced deposition rate and throughput of the equipment and the high cost of the collimator which must be removed and cleaned when its holes become clogged with the target material.

4.3.1   The Reactor Model

The reactor model for the magnetron sputtering must be able to describe the main factors which determine deposition uniformity. Since the sputtered molecules originate from the target surface, the transport mechanism to the substrate must be described. The sputter intensity of the molecules from the target depends on the properties of the plasma and the magnetic field. Simulation of this intensity requires a model of magnetically enhanced low pressure plasma. Furthermore a model for the interaction of high energy ions bombarding the target must exist. Fortunately, the sputter intensity is a measurable quantity, since during the process the target suffers a macroscopic erosion. A measurement of the erosion profile provides directly the intensity profile of the sputtered molecules. Moreover, there exist good models for the energy distribution of the sputtered molecules. Consequently, the reactor simulation only has to focus on the molecular transport through the ambient gas.

The model for the molecular transport assumes that the presence of ionized molecules and electric fields in the reactor does not influence the ambient properties significantly with respect to the sputtered molecules transport. This is justified because the plasma does not lead to a significant heating of the ionic or neutral gas component. Furthermore, the heating of the gas from accelerated ions in confined to the plasma sheath region, which is defined by the cyclotron orbits of the magnetic field and usually close to the target. In summary, it may be assumed that the ambient, for the purpose of sputtered molecule transport, is a gas.

The sputtered particles have a kinetic energy on the order of electron Volts up to the maximum energy of the DC-voltage, which is several hundred electron Volts. Compared to this energy scale, the thermalized ambient Argon atoms with an energy of about $0.025 \ eV$ are at rest. But it is only during the first few collisions that the sputtered molecules have this characteristic. After a few collisions, the sputtered molecules slow down so much that the velocity distribution of the ambient gas becomes significant. Then ordinary diffusion takes over. The trajectory of a sputtered molecule on the way to the substrate thus depends very much on the background pressure. For the magnetron processes at the lowest pressure below 1 $mTorr$, the mean free path is of the order of the target-substrate separation. This corresponds to a Knudsen number for the sputter particles of about 1.0. Collisionless transport and transport with one or two collisions dominates. In such cases the deposition rate can be obtained from a pure geometrical calculation (free molecular flow). For slightly higher pressures, free flow transport still dominates, so that the pure geometrical calculation gives an estimate, but the angular distributions are broadened. As the pressure increases further, the angular distribution of the sputtered molecules at the substrate still shows a forward characteristic, but the signature of the source is lost. This occurs around Knudsen number 0.1. The transition from a nearly free molecular flow to a diffusion dominated process occurs in the pressure interval $1-10\ mTorr$, which is precisely the operating range of the sputter processes. It is exactly in this regime that the effect of the background on the angular distribution of the molecules ejected from the target is most significant. Because in this pressure regime the angular distribution is very sensitive to the interaction with the background, a rather sophisticated model is necessary for accurate predictions. On one hand, the energy dependence of the collision cross section must be accounted for. The cross section of the highly energetic molecules is only half as large as the cross section of those molecules which have already undergone several collisions. On the other hand, the details of the angular scattering become important.

That there is significant interaction between the sputtering particles and the background gas can be seen from the so called sputtering wind effect [84]. The high kinetic energy of the sputtering particles is transferred to the background gas, which is heated up. The particle density becomes reduced and in consequence the collision rate and the energy transfer from sputtering particles to the background is reduced. This interaction can be significant, especially at higher pressures. The state in such a sputtering process can only be calculated with a self-consistent method.

The above discussion leads to the conclusion that a fully self-consistent DSMC simulation of the sputter process is necessary. The simulation should involve a two component mixture consisting of the background gas and the sputtering particles far from equilibrium. Such a calculation is automatically self-consistent; the only problem is that the concentration of the sputtering particles may be very low. If the mole fraction of one species is much below 0.1%, the results show a high statistical scatter. In the simulated reactor at a pressure around 0.13 Pa, the Titanium mole fraction was mainly above 0.1% and a reliable calculation of the state inside the reactor was possible. The statistical accuracy was also sufficient to calculate deposition rates.

But the desire to calculate angular distributions at different positions inside the reactor necessitated the use of the Test Particle method TP. In the TP method the movement of the sputtered molecules is calculated against a fixed background. Because the ambient gas molecules are absent from the simulation, it is possible to collect much better statistics on the sputtered molecules. In the simplest version of this method the ambient gas molecules have zero initial velocity before the scattering [68], [69], [94]. With this model, only the highly energetic scattering events can be accurately described. However, the diffusive behavior of the sputtered molecules which have already undergone several collisions is lost. In an improved version, the ambient gas molecules have a fixed velocity distribution which allows for diffusive transport [72]. In the investigation presented here, a further improvement was introduced. A hybrid model between a DSMC simulation and a TP simulation was created by taking the background not to be a gas with constant pressure and temperature, but a gas in a state which was calculated from a previous, self-consistent DSMC calculation. The background mass density in such a hybrid DSMC/TP-model will be reduced because of the sputtering wind effect and will then lead correctly to a reduced collision rate in the ambient. Also the increased background temperature in such a hybrid DSMC/TP-model will show correctly the effect of faster diffusion at elevated temperatures. Care must be taken as to how the temperature is defined for the background distribution. The concept of a temperature is closely related to a Maxwellian velocity distribution. One possibility is to use the average kinetic energy of the background component alone for the definition of a background gas Maxwellian. The validity of such an assumption will be discussed below.

As described in the last chapter, there are certain conditions which must be imposed on the numerical parameters of the calculation to ensure convergence. Here the primary concern is that the time step be chosen small enough to satisfy two conditions. First, a particle with the expected value of velocity magnitude should only travel some fraction of a cell per time step (maybe 3/4). Second, the fastest particles, say those with thermal speeds three times greater than the mean thermal speed, should travel less than a mean free path per time step. In our calculation, particles with an energy below 15 eV fulfilled these strict rules. An error in the propagation of the particles with higher energy is introduced only if their background changes significantly within one time step in the variables determining the collision rate.

The quantitative prediction of the molecular transport is further complicated by the fact that the molecular cross sections behave differently at high energy from the usual molecular model predictions like the Lennard-Jones model. This is because the highly energetic collisions probe more central parts of the molecular potential which are not captured in the Lennard-Jones parameters. The Lennard-Jones model has of course a velocity dependent total cross section, but this dependence is to weak for the high energies. A better approximation is given by the Abrahamson potential (see Section 2.4). It is possible to modify the temperature exponent (and thus the velocity dependence) of the VHS and M1 models to match velocity dependence associated with the Abrahamson potential.

A second issue concerning the numerical modeling of the collisions is the scattering behavior. The VHS model, which is usually employed in DSMC calculations, uses hard sphere scattering which leads to systematically too a low diffusion coefficient. This is of particular significance in high energy scattering. To remedy this situation, the M1 model was introduced in Section 2.4. In the simulations below, the VHS and M1 molecular models with the Lennard-Jones velocity dependence and the M1 model with the Abrahamsom velocity dependence were compared. Table 4.2 describes the parameters associated with these models.

Ar and Ti parameter	VHS	M1$_{LJ}$	M1$_{ABR}$
$\mu(273°K)$[Kg/ms]	$2.162 \ 10^{-5}$	$2.162 \ 10^{-5}$	$2.162 \ 10^{-5}$
ω	0.6785	0.6785	0.75
D $[m^2/s]$	2.057	2.594	2.527
deflection angle $\chi(b)$	hard sphere	linear	linear

Table 4.2: Parameters for the DSMC calculations

Finally the target model must be explained. The angular distribution was taken from a cosine law, and the energy distribution was taken from a Thompson distribution [93] as

$$\Phi(E,\theta)d\Omega dE = P\cos\theta \frac{1 - \sqrt{\frac{E_b+E}{\Lambda E_{inc}}}}{E^2(1 + E_b/E)^3}d\Omega dE \qquad (4.15)$$

$$\Lambda E_{inc} = \frac{4M_{ion}M_{target}}{(M_{ion} + M_{target})^2}E_{inc}$$

where E_{inc} is the average incident ion energy, ΛE_{inc} the maximum recoil energy, M_{ion} and M_{target} the ion and target species molecular masses, E_b the binding energy inside the target and P a normalization factor. We used $E_{inc} = 500 \ eV$, $E_b = 5 \ eV$.

This Thompson model has experimental confirmation [99]. The normalization factor P was adjusted in the following way: the measured erosion profiles on the target defined the position dependence of P. The overall normalization of P was chosen to provide the measured deposition rates.

4.3.2 Velocity Distributions

[1] In the description of the reactor model above, the effects of the sputtering wind, diffusive transport of the thermalized component, difference between various molecular models and the validity of a reduced description of the sputter process by the TP method were discussed.

These questions can be studied in a simplified system which attempts to model a single spatial dimensional as closely as possible. In this frame work, the effects mentioned above can be computed with DSMC with sufficiently high accuracy such that the velocity distribution function itself can be interpreted. The validity of the TP method can thus be tested in a numerical experiment. The redistribution of the sputtered particles within the reactor is a process which can be studied best in the more realistic

---

[1]Portions of this section are reprinted from [44] by courtesy of the American Institute of Physics

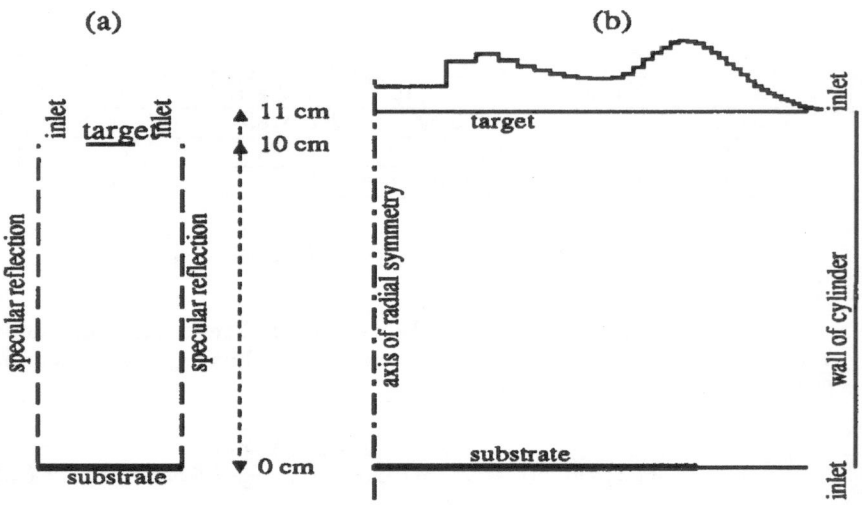

Figure 4.21: The quasi 1-dimensional and 3-dimensional reactor geometries.

3D axisymmetric geometry of a sputter reactor. The predictions from these models will be different for some effects at a higher pressure.

The Quasi One-Dimensional Sputtering Process

In this part the results of a quasi one-dimensional sputtering simulation are described. The starting point in this study of the nonequilibrium effects is the desire to study macroscopic variables and the distribution functions in a situation free from geometric effects. This suggests a 2-dimensional rectangular domain with specular reflection at the left and right walls representing homogeneity in the lateral direction and with a target and substrate at the top and bottom. This allows the demonstration of the sputtering wind effect, i.e. the density reduction in the gas through the heating from the sputtered molecules. The degree of density reduction can only be specified when there is a reference state with no sputtering. This necessitates the introduction of inlets in order to establish a connection to a gas state without sputtering and a well defined reference pressure. Figure 4.21 (a) shows a sketch of this geometry.

This setup will be called 'quasi one-dimensional' because it is the closest possible approach to a situation homogeneous in the lateral direction. The connection of the sputtering chamber to a region 'without sputtering' is also a good approximation to the experimental situation, where the pressure gauge is removed from the sputtering chamber and is surrounded by a gas close to equilibrium and at wall temperature. It also includes the effects

Table 4.3: Sputtering rates from target, deposition rates for $M1_{LJ}$ model.

Pressure [mTorr]	sputtering rate $[10^{20}/m^2 s]$	deposition rate $[nm/s]$	mfp at 0.1eV [cm]	mfp at 10eV [cm]
0.5	2.333	2.15	13.3	36.9
1.5	7.000	5.89	4.43	12.3
5	23.33	17.1	1.33	3.69
5	3.333	2.21	1.33	3.69

of a finite residence time, i.e. the gas transport through inlet and outlet.

To observe realistic effects in the simulation, it is important to choose an adequate sputtering rate. In Table 4.3 the sputtering rates and associated deposition rates for the three pressures considered are collected. At the highest reference pressure of $P = 5\ mTorr$ two different rates are used. The higher rate is probably above the achievable level, so that the observable range lies between the results of the two rates. The table also contains a reference mean free path λ calculated from $\lambda = 1/(\pi d_{ref}^2 n_{ref})$ where n_{ref} is the reference particle density.

Figure 4.22 shows the mass densities of the background gas as a function of the vertical position in the quasi one-dimensional geometry for the sputtering rates described in Table 4.3. The target to substrate distance is $0.1\ m$. These are compared with the reference density, i.e. the mass density defined through the particle density and temperature boundary conditions at the inlets. For the highest pressure case Figure 4.22 also shows the effect of the three tested scattering models. The decrease in density has a strong pressure dependence corresponding to the experimental findings of Rossnagel [84]. Whereas at 0.5 $mTorr = 0.0658\ Pa$ there is almost no density reduction effect, at 5 $mTorr = 0.658\ Pa$ it becomes a 20% – 50% effect, depending on sputtering rate and model. Moreover, in spite of the different geometry of Rossnagel's reactor, the same distance between the target and the density minimum, namely 5 cm, is found, independent of pressure. This is the position where the energy transfer from the sputtered atoms to the background gas is maximal.

Also strong is the dependence of this effect on the scattering model. With the $M1_{ABR}$ model and its smaller cross sections at high energy, there is less energy transfer into the gas than with the $M1_{LJ}$ model with its relatively large cross sections at high energy.

Figure 4.23 contains plots of the Ti-concentration dependence on vertical position in the 1-D geometry. The figure contains the concentrations for the pressures 0.5 $mTorr$, 1.5 $mTorr$, and 5 $mTorr$ with a sputter rate of

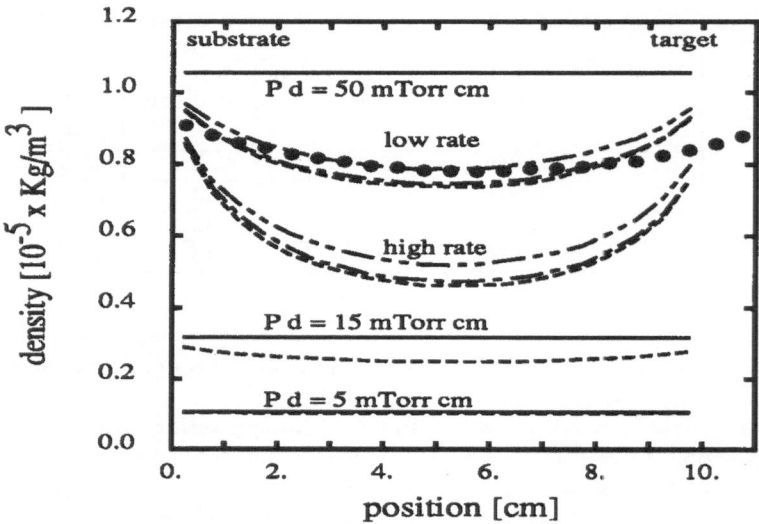

Figure 4.22: Mass density of the background gas in the 1-dimensional reactor. The reference density (—), the density with the M1(LJ) model (- - -), with the M1(ABR) model (-..-..) and with the VHS model (-.-.) are shown. The dots give the density in the 3-D geometry.

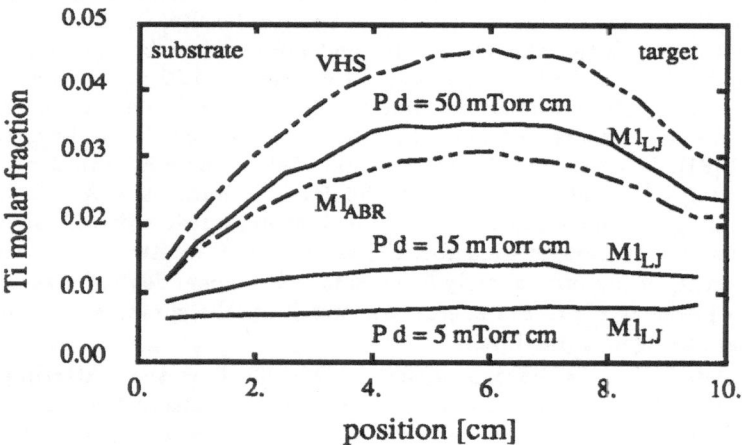

Figure 4.23: Titanium molar fraction in the 1-D reactor. The results for the different scattering models at 50 *mTorr cm* show the effect of the different diffusion coefficients.

23.33×10^{20} $1/m^2s$, as well as results for the three molecular models at the highest pressure. The maximum of the concentration depends on the value of the diffusion cross section $Q^{(1)}$ (see Equation 2.59). Model M1, which has a higher diffusion coefficient than the VHS model, shows a significantly reduced maximum value.

More information about the state in the sputtering chamber is contained in the energy distributions of the Ar and Ti components. Figure 4.24 shows several such distributions at the substrate position in the 1-dimensional geometry. They result from sampling the particles traversing the first cell in front of the substrate during DSMC calculations (with the $M1_{LJ}$ model). The figure also contains an equilibrium distribution for a temperature of $300° K$ as a reference.

The first set of results show Ar and Ti distributions for the low sputter rate of 3.33×10^{20} $1/m^2s$ at 5 $mTorr$. The Ar distribution is similar to an equilibrium distribution, but is centered at an increased temperature of $\log E = -1.4 \equiv 500° K$. The Ti distribution has most of its intensity at low energy (the 'thermalized' component), but also a significant tail of high energetic particles. Surprisingly, the low energetic part of the distribution corresponds to an equilibrium distribution with a temperature of about $\log E = -1.2 \equiv 750° K$, which is much higher than the background temperature.

The second set of results show Ar and Ti distributions for the high sputter rate of 23.33×10^{20} $1/m^2s$ at 5 $mTorr$. Again the Ar distribution is similar to an equilibrium distribution at a higher temperature, but now there is a significant high energy tail. The Ti distribution, on the other hand, no longer resembles an equilibrium distribution. It is not possible to distinguish a 'thermalized component' and a 'high energy tail'. The maxima of the Ar and Ti distributions are very different. It follows that the energy distribution of the background gas atoms cannot be considered to be in a state close to equilibrium. Moreover, the maximum of the Ti distribution is clearly separated from the peak of the Ar distribution.

That the nonequilibrium character of the Ar background gas is responsible for the observed thermalization effects of the Ti atoms can be seen in Figure 4.25. It contains Ti energy distributions, but now calculated with the TP method and with different assumptions about the background distributions. The first choice is an equilibrium distribution at $\log E = -1.6 \equiv 300° K$. The second choice is an equilibrium distribution at $\log E = -1.4 \equiv 500° K$. This temperature comes from the position of the maximum of the Ar distribution of the DSMC calculation shown in Figure 4.24. The third choice is an equilibrium distribution at $\log E = 0.2 \equiv 19000° K$. This corresponds to the average kinetic energy of the Ar atoms in the DSMC calculation. For comparison, Figure 4.25 also contains the Ti energy distribution from the DSMC calculation (the same as in Figure 4.24) at the low and high sputter rates. The first observation

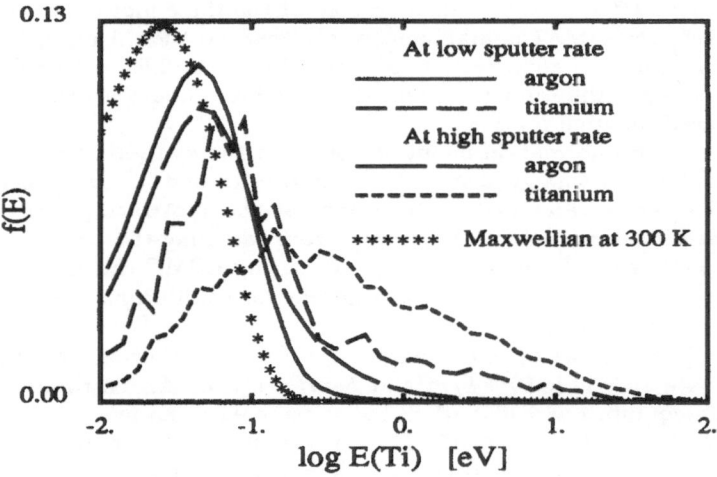

Figure 4.24: Ti and Ar energy distributions at the substrate from DSMC calculations in the 1-D geometry.

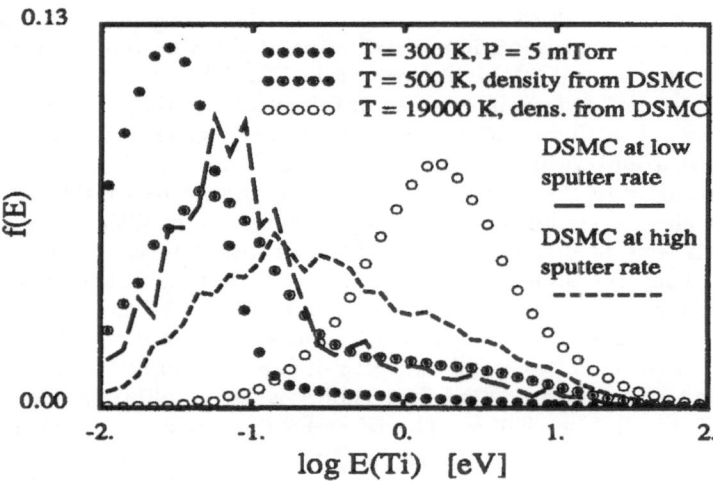

Figure 4.25: Ti energy distributions at the substrate from TP calculations in the 1-D geometry (dots) compared with the results from the self-consistent DSMC calculations (lineplots).

is that all distributions from the TP calculation have a maximum exactly at the position of the maximum of the background equilibrium distribution plus a high energy tail. It is now clear that if the maximum of the Ti distribution deviates from the maximum of the Ar distribution, as was the case in the DSMC calculations, this background distribution is not in equilibrium. The second observation is that no choice of an equilibrium distribution for the background can provide a Ti distribution matching the true Ti distribution calculated with the DSMC method. Clearly, a choice of $T = 500° K$ in the TP calculation, which neglects completely the highly energetic part of the background, leads to a Ti distribution which is 'too cold', whereas a choice of $T = 19000° K$ overestimates the influence of the high energy tail on the Ti thermalization. The optimal choice is somewhere in between. Since the TP calculations do not depend on a sputter flux, this optimal choice must also take this into account.

The result suggests that for the case of a high sputter flux, only the DSMC calculation gives valid results for the Ti energy distribution, whereas for the case of a low sputter flux, the TP calculation gives a good approximation for the Ti energy distribution, if an estimate for an Ar background temperature is given. A good choice for this temperature can be obtained by fitting an equilibrium distribution to the maximum of the Ar distribution from a DSMC calculation (in our calculation it was $\log E = -1.4 \equiv 500° K$).

The Three Dimensional Sputtering Process

The main application of sputtering simulation is to the optimization of process conditions and reactor geometry of realistic production reactors. This requires simulations in a 3-D geometry. In this section we show that the effects studied in the quasi 1-D situation are also present in the 3D-situation. Figure 4.21 (b) shows the geometry of a 3D axisymmetric reactor which resembles existing equipment. The connection to a region of reference pressure is established with two inlets at the top and bottom. The erosion profile plotted above the target gives the normalized local sputter rate. It comes from a measurement of the erosion profile of a reactor. Since the emission of sputtered atoms is assumed to follow the cosine law, no corrections must be made to first order to take the geometric effects of erosion (i.e. the fact that the normal of the target surface is no longer perpendicular to the target plane) into account. The depth of the erosion profile is on the order of a few millimeters. The average sputter rate is 3.33×10^{20} $1/m^2 s$, and the reference pressure is 0.7 Pa. In a 3-D geometry one expects a reduction in the magnitude of sputtering.effects, as the gas can also lose energy at the walls. This is confirmed by the density reduction plotted in Figure 4.22 (dotted).

The Ti and Ar energy distributions are plotted in Figure 4.26. They are similar to the distributions in the quasi 1-D case. Again, the maxima of the Ar and Ti distributions are different.

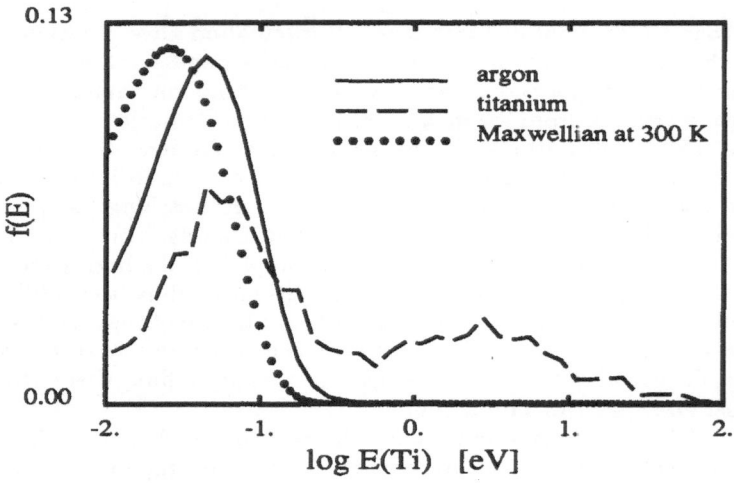

Figure 4.26: Ti and Ar energy distributions at the substrate from a DSMC calculation in the 3D geometry.

For reactor optimization the question of key importance concerns the deposition rates of sputter atoms on the substrate and target. Figures 4.27 illustrates the effect of the sputtering wind on the titanium deposition rates for both the low and high pressure cases by comparing the fully self-consistent DSMC calculation with its nonequilibrium background gas to the TP method with the background gas at $300°K$ and at reference pressure. At low pressure, both TP and DSMC give almost the same result. At the higher pressure, the difference of the results is significant, especially for the backscattering on the target, which is shown in Figure 4.28.

Conclusions

In this investigation, the influence of various model assumptions on the validity of Monte Carlo simulations for a Ti sputter process was studied. The main results can be summarized as follows:

(1) The influence of the Ti sputter atoms on the Ar background gas leads to a reduced density and nonequilibrium energy distribution of the Ar atoms. An accurate simulation requires a self-consistent treatment of both sputter atoms and background gas with the DSMC method whenever the condition $Pd \geq 5 \; mTorr \; cm$ is fulfilled, where P is the pressure and d the distance between target and substrate.

(2) At moderate pressures the Test Particle method can be used, which calculates the Ti transport in a background defined by a position dependent

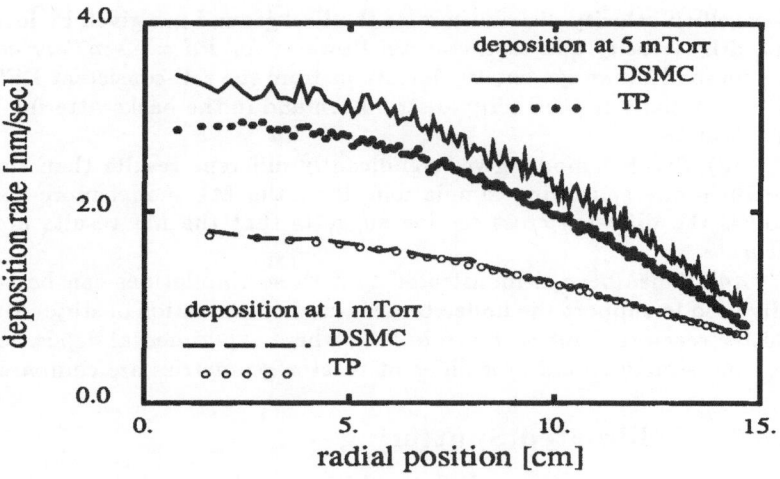

Figure 4.27: Deposition rate on the substrate in the 3D geometry.

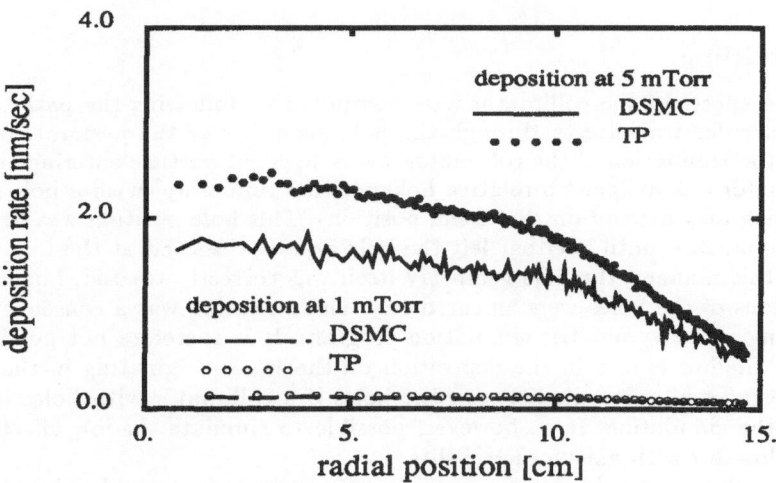

Figure 4.28: Deposition rate on the target in the 3-D geometry.

Maxwellian velocity distribution for the background gas given by local density, drift velocity and temperature. However for $Pd = 50 \; mTorr \; cm$ even this method shows systematic deviations from the self-consistent DSMC results, especially in the energy distribution and in the backscattering rate of the Ti atoms.

(3) The M1 model gives significantly different results than the VHS model in the sputtering simulation. That the M1 model more correctly models the diffusion cross section suggests that the M1 results are more accurate.

(4) It has been demonstrated that these simulations can be successfully used to support the understanding and optimization of state-of-the-art sputter reactors. This is shown in [45], where experimental deposition profiles and simulated data for different reactor geometries are compared.

4.3.3   Collimated Sputtering

A collimator is a perforated plate which is positioned between the target and the substrate. While the collimator holes allow sputter particles with a small normal (to the collimator surface) angle to pass through the hole (the maximum angle depends on the ratio of the height of the hole to its diameter H/D), the sputter particles with large normal angles will be absorbed inside the holes.

Modeling

The effects of the collimator were computed by following the paths of the molecular trajectories through the hole geometry of the perforated plate. In the simulation of the collimator, every incident particle entering the collimator was assigned a relative hole position randomly (with a hole radius which may depend on the radial position). This hole position was fixed for this particle until it either left the hole or was absorbed at the inner wall. In this manner, the hole geometry itself was correctly treated, but the positions of the holes were smeared out. This smearing was a consequence of using an axisymmetric simulation program. It is therefore not possible to see shadow effects in the deposition on the wafer originating in the finite thickness of the walls between the holes if a collimator with holes is used in the simulation. It is, however, possible to simulate shadow effects of a collimator with axisymmetric slits.

The modeled reactor is cylindrically symmetric, has a height of $11 \; cm$ from wafer to target and has a radius of $15 \; cm$. The target has a radius of $14 \; cm$. There is a gas flow of the Argon background gas from an inlet, ring shaped at the bottom below the wafer, to an outlet, also ring shaped, at the top above the target area. The gas flow influences the simulation results very little.

The collimator is a perforated plate with radius $15\,cm$ located between the target and the substrate It has holes covering the whole area. The collimator is defined by its thickness H, by its hole diameter D, by the fraction of the area of holes to the total area A, and by the distance X of the lower side of the collimator surface to the wafer.

With this collimator model, the $M1_{Abr}$ molecular model, the DSMC method and a Titanium flux from a measured erosion profile, a simulation of the collimated sputter process was undertaken.

Figure 4.29(a) shows the molar concentration of Titanium in the presence of the collimator plate. Since only those Ti atoms with a very small angle normal to the plate can pass through the collimator, their density below the collimator is decreased by a factor of 10 in that case. However, their angular distribution is much better suited for good edge coverage.

Figure 4.29(c) shows the temperature distribution of the argon carrier gas calculated from the mean kinetic energy in this nonequilibrium situation. Due to the energy transfer from the high energy Ti atoms, the carrier gas is heated above the collimator (this is the sputtering wind effect). This effect is visible, although this process employs the lowest possible pressure. Additional inhomogeneities below the collimator arise from heating the substrate.

The scattering rate of the Ti atoms is mainly influenced by the mass density distribution in the background gas, which is shown in Fig.4.29(b). In the middle of the reactor the interaction with the sputtered Ti atoms reduces the density by 5%. Below the collimator the heating from the substrate reduces the density significantly. Note that all previous studies have neglected these effects by assuming a homogeneous background density.

Design Variations

As mentioned above, the collimator has a negative impact on the overall deposition rate since only a small fraction of atoms can pass through the collimator. This can be studied in Figure 4.30, where deposition rate profiles are depicted both without collimator and for collimators with different openings.

In the next pictures the dependence of the deposition rate on the collimator and erosion profile is studied. Figure 4.31(a) shows the deposition rate for the different titanium flux distributions from Figure 4.31(b), but the same collimator with $H = 1\,cm$, $D = 0.6666\,cm$ and distance $X = 3\,cm$ from the lower side of the collimator to the substrate. The increase of the erosion profile increases the deposition in the central region, but not in the outer regions of the wafer. It is obvious that a good measurement of the erosion profile is necessary to provide a simulation which is consistent with the deposition measurements.

Figure 4.32 shows the deposition rate for different openings of the

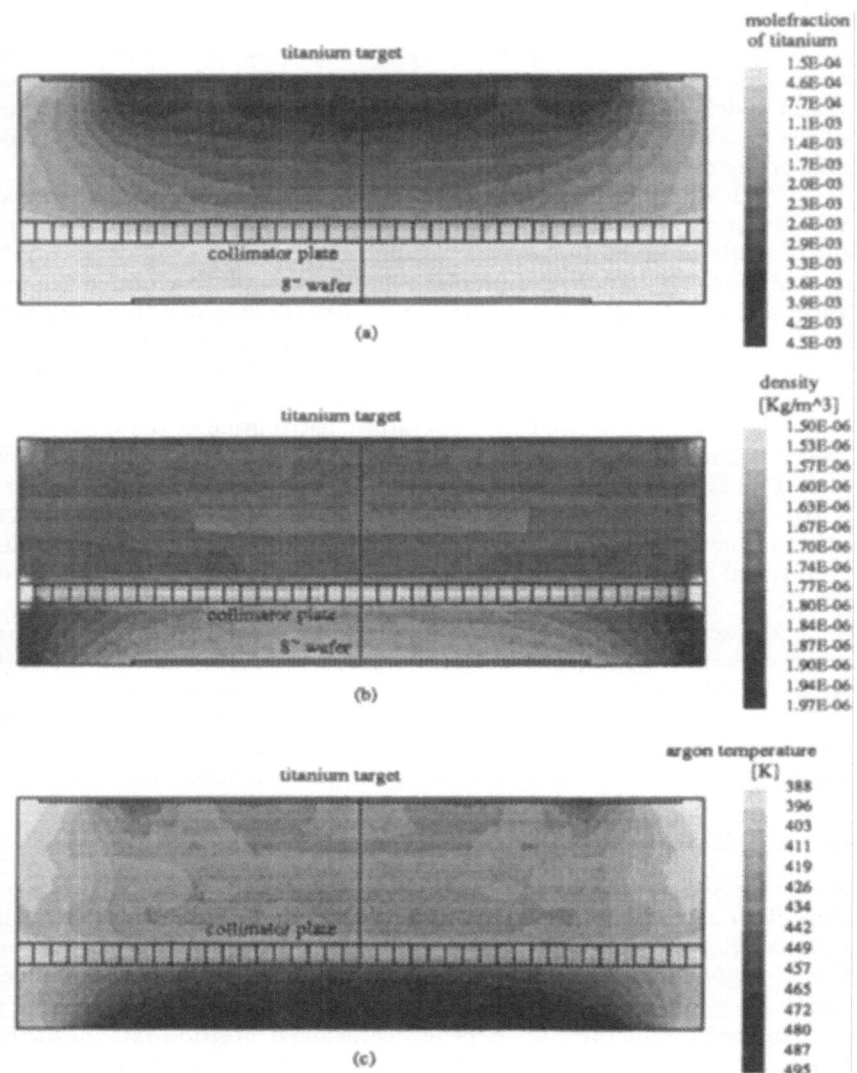

Figure 4.29: Figure (a) shows the molar concentration of titanium inside the reactor. The Ti is depleted behind the collimator. Figure (b) shows the mass density in the reactor. Above the wafer, the density reduction results from the sputtering wind effect. The density reduction below the collimator comes from the heating of the wafer to $500°K$. Figure (c) shows the the argon temperature defined as the mean kinetic energy. This temperature has a local maximum at the target.

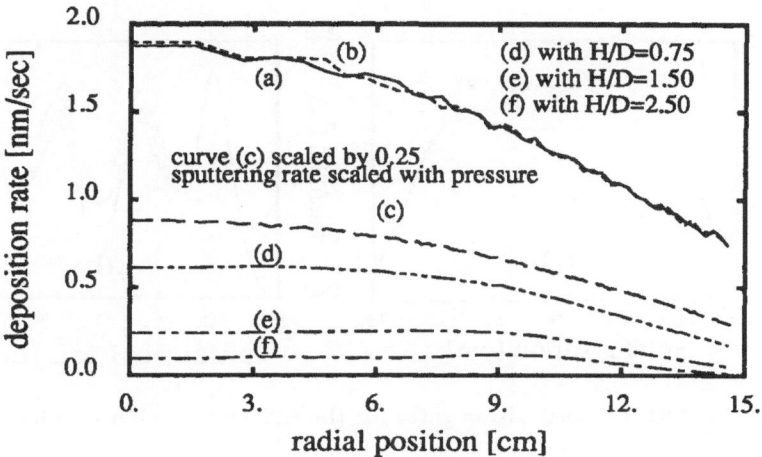

Figure 4.30: Deposition rate profiles along a radius across the substrate for different working conditions: a)-c) without collimator: a) 0.13 Pa, $T_{sub} = 250°C$, b) 0.13 Pa, $T_{sub} = 50°C$, c) 0.7 Pa, $T_{sub} = 50°C$ d)-f) collimators with different H/D, 0.13 Pa, $T_{sub} = 250°C$.

collimator with same thickness $H = 1\,cm$ and distance X. For very large aspect ratios the deposition rate on the substrate is a mapping of the sputter intensity from the target above.

Figure 4.33 shows deposition rates for collimators with the same ratio $H/D = 1.5$, but different positions and thicknesses H. The differences in the deposition rates are very small. Consequently, once an erosion profile and a collimator are found which produce acceptable uniformity, H and D can be increased if H/D remains constant to increase the lifetime of the collimator. A larger hole radius increases the lifetime of the collimator, because with the same deposition rate of titanium at the inside walls, the hole cross section decreases more slowly with larger holes than with smaller holes. The holes of the collimator should thus be made as large as possible keeping in mind the fact that the distance X should be large enough to avoid shadow effects.

The next results concern the aging of the collimator, i.e. the filling and closure of the collimator holes with titanium with time. This aging necessitates the removal of the collimator and thus constitutes a significant cost factor.

Figure 4.34 shows the behavior of the deposition rate when the hole radius shrinks inhomogeneously according to the inhomogeneous deposition inside the holes. The aging of the collimator is taken into account in the simulation in the following way: When a particle is absorbed inside a hole,

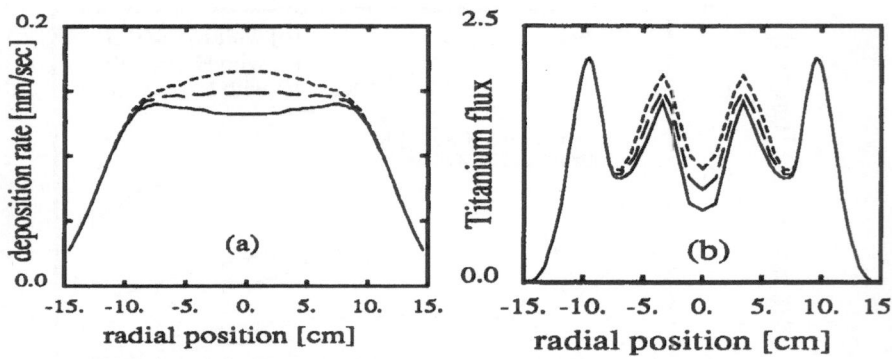

Figure 4.31: (a) Deposition rates for the different erosion profiles of (b)

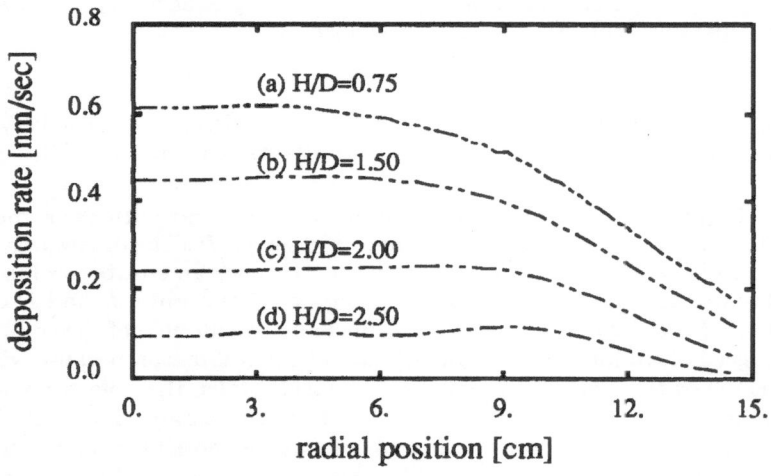

Figure 4.32: Deposition rates for different collimator openings. Collimator with same thickness and position. Curves (a),(c),(d) identical with (d),(e),(f) from Fig. 4.30. As the aspect ratio increases, the deposition rate resembles more and more the erosion profile on the target.

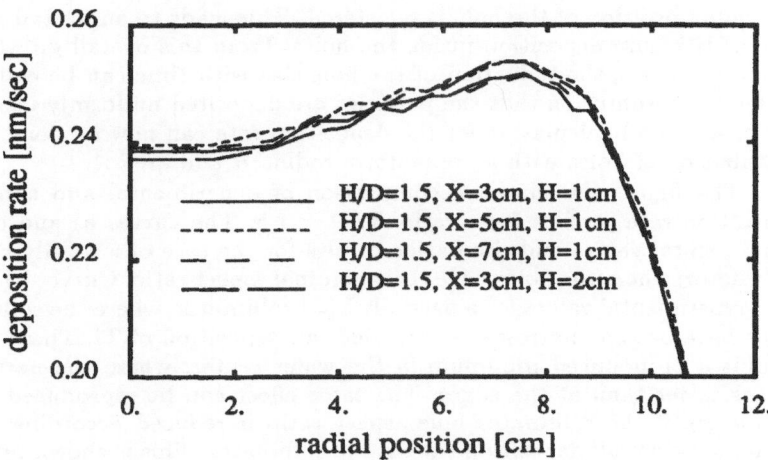

Figure 4.33: Deposition rate for collimators with the same ratio H/D=1.5, but with different opening, thickness and position.

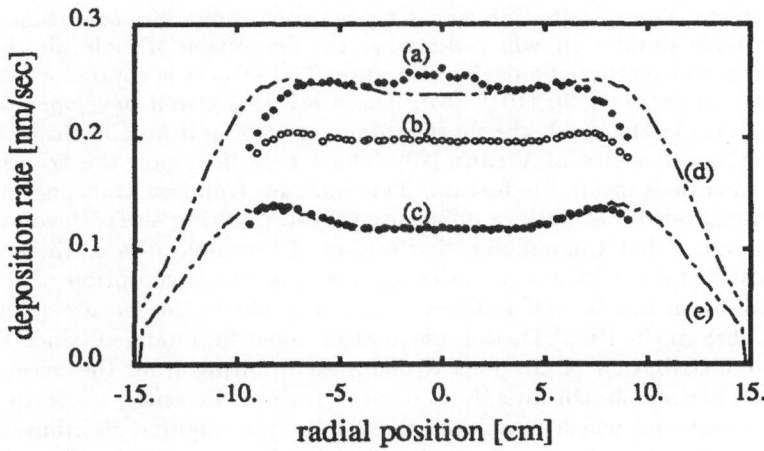

Figure 4.34: Experimental (dots) and simulated (lines) deposition rate profiles along a diameter across the substrate for a virgin and two partially aged collimators.

the radial position of the hole is registered. This leads to an overall growth rate of titanium deposition inside the holes. From this overall growth rate inside the holes, the reduction of the hole size with time can be estimated under the assumption that the particles are deposited uniformly inside the holes. A second calculation for the deposition rate can now be done with a distribution of holes with a nonuniform radius R and area A.

The figure also shows a comparison of experimental and simulated deposition rate profiles for a ratio $H/D = 1.5$. The curves a) and d) give experimental values and simulation results for the case of a freshly cleaned collimator, where the holes have their nominal aspect ratio. Curves b) and c) give experimental values for a partially aged collimator, where the collimator holes have become narrower due to sidewall deposition of Ti. These curves exhibit a pronounced minimum in the wafer center, where the narrowing occurs faster than at the edges. The same effect can be reproduced in the simulation if the collimator hole aspect ratio is reduced according to the simulated sidewall deposition rate in the collimator. This is shown by curve e) in Figure 4.30.

Angular Distribution

One of the most useful results that can be obtained from a sputtering simulation is the angular distribution of the sputtered atoms arriving at the substrate. This distribution must be given as the boundary condition for feature scale simulation, which describes the deposition of molecules inside a microscopic feature of a device structure. The subject is covered extensively in the literature [82], [101], [61]. There are many well developed simulation tools in this field, the most prominent coming from CIS Stanford [62] and the University of Vienna [89]. These tools focus on the transport of the molecules inside the feature. This includes Knudsen transport, surface chemical reactions, surface diffusion and the resulting surface evolution. It is assumed that the angular distribution of the molecules at the entrance of the feature is given. In CVD applications, the assumption of a cosine distribution law is well fulfilled, as the gas above the feature is close to equilibrium. In PECVD, this assumption must be modified, since the angular distribution of the ions is distorted resulting from the acceleration in the plasma sheath directly above the feature. Recently, a sheath model was developed which is able to predict the ion angular distribution [19]. For sputter deposition the angular distribution deviates significantly from an equilibrium distribution. Often, for lack of more complete information, either a parametrized form of the angular distribution is assumed or the results of a simplified sputter simulation without collisions in the gas phase are used. The results of the simulation presented here illustrate, however, the importance of including the interaction with the background gas.

Depending on the application, various angular distributions may be of interest. For example, the feature simulation of a contact hole requires the full angular distribution. This is computed as a function of the steric angle. The value at a given angle gives the flux intensity in that direction. The graph of this function is given by a deformed surface of a sphere. For the feature simulation of trenches, however, it is only necessary to compute the angular distribution projected onto a plane which is orthogonal to the direction along the trench and to the substrate itself. This angular distribution assumes a translational symmetry (orthogonal to the projection plane).

To determine the influence of the background gas on the angular distribution of the molecules above the wafer, a test particle simulation was done. To accurately evaluated the distribution function, it was necessary to use full 3-D axisymmetric geometry. This precluded the use of the full DSMC method due to the enormous amount of computation time that would be required. The first two simulations modeled titanium sputter equipment without a collimator. The pressure in the first simulation was 0.13 Pa, which is at the low end of the achievable pressure in a magnetron. For the second simulation the pressure was 0.8 Pa, which is at the high end of operating pressures with a magnetron. The plane projected angular distribution was recorded at various radial positions directly above the wafer. Figure 4.35(a) shows the results from the sputtering at low pressure. The distribution shows a strong dependence on the radial position because the molecules have not changed their direction very much on the path from the target to the substrate. This result is close to a simulation without interaction with the background. Figure 4.35(b) shows the results from the sputtering at higher pressure. The dependence of the distribution on the radial position is almost lost. The distribution still has a forward characteristic, but tends to a more or less uniform distribution as in the case of diffusive transport. The last simulation was done for a collimated process at 0.13 Pa. Figure 4.35(c) shows the angular distribution in several radial positions. The distribution has the desired strong forward characteristic. The price to be paid for this, however, is that of a reduced deposition rate.

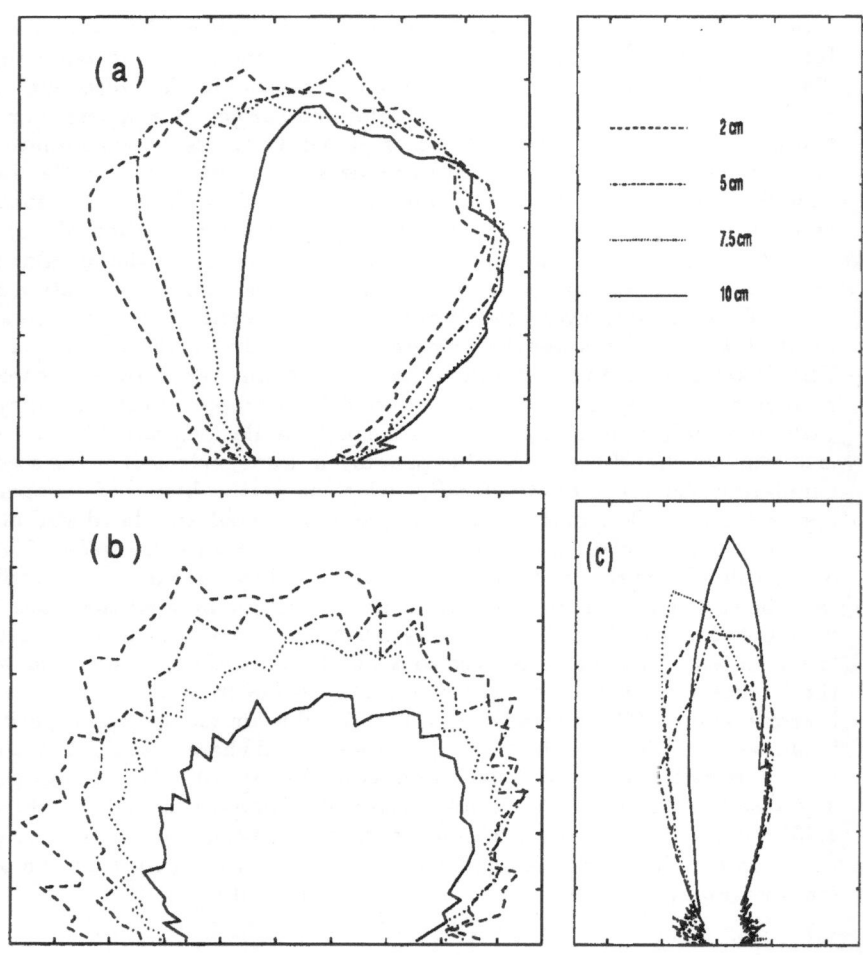

Figure 4.35: Plane projected angular distribution of titanium above the wafer in several radial positions. The sputter equipment has no collimator in (a) and (b) and a background pressure of 0.13 Pa in (a) and 0.8 Pa in (b). In Figure (c), the sputter equipment has a collimator, and the background pressure is 0.13 Pa.

Chapter 5

Modeling of Radiative Heat Transfer

We now continue with the topic of radiation transport first introduced in Section 1.3. After a brief review of applications of radiative heat transfer in micro-electronics manufacturing equipment, we present a more detailed look at radiation scattering in semi-transparent materials. The results of Chapter 1 for scattering off opaque surfaces are extended to semi-transparent materials through the use of effective optical properties. In particular this allows an effective description of solids coated with thin film layers. Such surfaces appear frequently in micro-electronics processing. The dependence of general optical properties of surfaces on temperature and wavelength of radiation is also addressed.

Next we present a formal solution to the radiation transport equation as an infinite series obtained through iteration. It is this form which is evaluated using the Monte Carlo methods described in the next chapter. We also consider the simplifications of the solution obtained in the special cases of specularly and diffusely reflecting surfaces.

Finally the rendering equation, a modification of the radiative transport equation which is used in computer graphics, is discussed.

5.1 Rapid Thermal Processing

Thermally activated chemical vapor deposition is one of the major techniques for the fabrication of advanced ultra large scale integrated (ULSI) semiconductor devices [36]. Radiative heat transfer is therefore an important physical phenomenon in a large variety of existing processing equipment.

Up to now, the processing environment has been dominated by the diffusion furnace equipment. In such a furnace, a batch of wafers is processed in the cylindrical tube of a hot wall reactor, where severe temperature gradients during the ramp up to the processing temperature can only be held small by slowing down this process. These ramp times, however, are becoming unacceptably long for two reasons. First, wafer size is increasing as a means of reducing cost, and larger wafers require longer ramp times. Second, the inclusion of smaller, submicron devices on a wafer requires that the temperature budget be reduced. The temperature budget is a measure of the extent to which an unwanted process or reaction has occurred. Such processes may be more easily controlled through shorter ramp times.

More attention has been paid recently to the single wafer kind of equipment, which may enter mass production in the near future. Such equipment is known as Rapid Thermal Processing (RTP) equipment, or in the case of thermally activated deposition, Rapid Thermal Chemical Vapor Deposition (RTCVD) equipment. Typically, a single wafer is heated through a quartz window by tungsten halogen lamps. The chamber is either cold wall polished steel or additionally equipped with a chemically more inert quartz inlet which may, however, warm up and represent a warm wall system. The low thermal mass of the single wafer and the cold wall allow a rapid cycle time and minimal thermal budget.

Radiative heating of a large wafer in the presence of strong temperature gradients in the vicinity of the cold walls and in fulfillment of the strict uniformity requirements (see SRC roadmap in RTP [30]) is a nontrivial technological problem. Similarly, a simulator which is able to give reliable predictions for design and optimization must go beyond the traditional engineering methods for describing heat transfer in complex environments [18], [87]. Such a simulator must on one hand describe radiation transport on a relatively fundamental level such that the properties of the radiation source, the geometry and the material can be accurately accounted for, while on the other hand be numerically efficient, so that the results can be obtained within a reasonable amount of computation time. Furthermore, the radiative simulation must be coupled to a CVD simulator which solves for flow, mass transport and chemistry. This coupling is address in Section 6.3.

5.2   Semi-transparent Materials

Most of the interior of an RTP reactor is occupied by gas, which for the purposes of radiative heat transfer may be treated as a vacuum. The walls of the reactor are steel and thus may be treated as opaque surfaces. The modeling of scattering off such opaque surfaces is straightforward and was discussed in the first chapter. Usually, however, some portion of the reactor interior consists of semi-transparent materials. These are materials which allow transmission of radiation as well as reflection and absorption. Examples are the wafer itself, thin material layers on a silicon substrate, and quartz windows or enclosures. These semi-transparent regions require closer attention.

In some cases the semi-transparent regions may be handled purely through surface scattering, as opposed to following detailed ray paths through the material. When this is possible, a simplified version of the radiation transport equation may be applied, whereby the radiation is transported from surface element to surface element instead from volume element to volume element. This surface approximation may be made when the geometry of the semi-transparent material is simple enough - e.g. plane parallel. In this case the material properties may be integrated a priori to obtain effective values of reflectivity, absorptivity, emissivity and transmissivity. This is known as the McMahon approximation.

The idea of finding a particular solution to the radiation transport problem for regular geometries and expressing it in terms of effective optical properties can be generalized. Often, thin material layers show interference effects. To describe these effects within plane parallel geometries, the phase information of the radiation is necessary. This goes beyond Equation (1.92). But this phase information is only needed in microscopic dimensions. So it is possible to calculate the thin film effects separately and express them as effective optical properties. In the case of interference effects, these properties are given by the Fresnel formulas 5.2. Another example is the influence of microscopic structure or surface roughness on the optical properties. Models of surface roughness allow radiation transport to the nonplanar surface to be expressed in form of modified optical properties [59], [26], [37].

5.2.1 The McMahon Approximation

In the case of the McMahon approximation, Equation (1.92) can be solved explicitly for plane parallel geometry, and the sources and sinks of the radiation can be formulated to be at the surfaces of the solid as in the case of an opaque solid. The radiation transport is then purely surface to surface radiation. The assumptions are that the semi-transparent material is homogeneous and in particular that the absorptivity $\alpha_{\nu,T}(x) = \alpha_{\nu,T}$ is constant. Constant absorptivity implies that the transmissivity $t(x,y)$ between two points x and y depends only on the distance traveled through the material, but not the direction, i.e. $t(y,x) = t(x,y) = e^{-|x-y|\alpha_{\nu,T}}$. Any end effects do to the finite length of the parallel sides are ignored. Furthermore, it is assumed that the reflection and transmission are specular. Let $\mathcal{R}$ be the specular reflection operator and $\mathcal{S}$ be the operator describing the refraction according to Snell's law (5.1) from a dielectric medium with refractive index $n_{\nu,T}$ into a dielectric medium with refractive index $n'_{\nu,T}$.

$$n_{\nu,T} \sin \theta_1 = n'_{\nu,T} \sin \theta_2 \tag{5.1}$$

θ_1 and θ_2 are the angles of ω and $\mathcal{S}(\omega)$ relative to the surface normal (see Figure 5.1). ω is the unit vector describing the direction of propagation.

If the refractive indices of the two media were equal (and consequently θ_1 and θ_1 were equal), then there would be no reflection at the interface. However, when the refractive indices are different, the interface acts as a partially reflecting surface with a reflectivity which is a function of the difference in the angles. This reflectivity at a surface with a jump in the index of refraction, assuming specular reflection, is given by the Fresnel formula

$$r_{\nu,T}(x,\omega) = \frac{1}{2}\left(\frac{\tan^2(\theta_1-\theta_2)}{\tan^2(\theta_1+\theta_2)} + \frac{\sin^2(\theta_1-\theta_2)}{\sin^2(\theta_1+\theta_2)}\right). \tag{5.2}$$

Note that if a ray passes through a medium with refractive index $n_{\nu,T}$ into a medium with refractive index $n'_{\nu,T}$ then back into the first medium, by the symmetry of Equations (5.1) and (5.2), the ray must enter and leave the middle medium with the same angle θ_1. Moreover, all reflectivities at the two interfaces must be equal, i.e.

$$r_{\nu,T}(y,\mathcal{R}\mathcal{S}(\omega)) = r_{\nu,T}(x,\mathcal{S}(\omega)) = r_{\nu,T}(x,\omega) = r . \tag{5.3}$$

And as mentioned above, the transmissivity does not depend on the direction the ray travels in the semi-transparent material. Thus all transmissivities t in the problem are equal (for a given θ_1).

We first calculate the radiance transmitted through the plate as a function of radiance incident from the other side of the plate as shown in Figure 5.1 (a). The radiance emitted from the volume of the plate is considered separately below. The following relations hold

$$i_\nu^{(t)}(x,\omega) = \frac{|n(x)\cdot\mathcal{S}(\omega)|}{|n(x)\cdot\omega|}\left(1-r\right)t\, i_\nu^{(o)}(y,\mathcal{S}(\omega)) \tag{5.4}$$

$$i_\nu^{(o)}(y,\mathcal{S}(\omega)) = r\,t\, i_\nu^{(o)}(x,\mathcal{R}\mathcal{S}(\omega)) + \frac{|n(y)\cdot\omega|}{|n(y)\cdot\mathcal{S}(\omega)|}\left(1-r\right)i_\nu^{(i)}(y,\omega) \tag{5.5}$$

$$i_\nu^{(o)}(x,\mathcal{R}\mathcal{S}(\omega)) = r\,t\, i_\nu^{(o)}(y,\mathcal{S}(\omega)) . \tag{5.6}$$

From Equations (5.5) and (5.6) we have

$$i_\nu^{(o)}(y,\mathcal{S}(\omega)) = \frac{1}{1-t^2\,r^2}\frac{|n(y)\cdot\omega|}{|n(y)\cdot\mathcal{S}(\omega)|}\left(1-t\right)i_\nu^{(i)}(y,\omega) . \tag{5.7}$$

Substitution of (5.7) into Equation (5.4) yields the transmitted radiance as a function if the incident radiation

$$i_\nu^{(t)}(x,\omega) = t\,\frac{(1-t)^2}{1-t^2\,r^2}\,i_\nu^{(i)}(y,\omega) \tag{5.8}$$

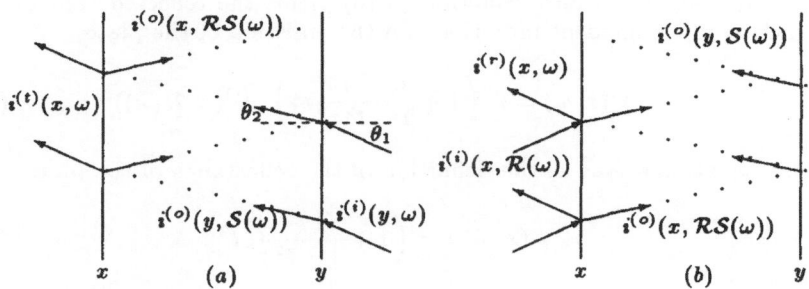

Figure 5.1: Transmission and reflection in plane parallel geometry. In (a), radiance transmitted through the plate to x is calculated as a function of the radiance incident on the plate at y. In (b), radiance reflected at x is calculated as a function of radiance incident at x and radiance emitted by the plate itself.

This equation serves to define the transmittance of the plate as

$$t^*_{\nu,T}(x,\omega) = t\,\frac{(1-t)^2}{1-t^2\,r^2} \tag{5.9}$$

We now consider the radiance reflected by the plate as a function of radiance incident from the same side of the plate and the radiance emitted by the plate itself as shown in Figure 5.1 (b). From the picture it can be seen that

$$i^{(r)}_{\nu}(x,\omega) = \frac{|n(x)\cdot\mathcal{S}(\omega)|}{|n(x)\cdot\omega|}\left(1-r\right)i^{(o)}_{\nu}(y,\mathcal{S}(\omega)) + r\,i^{(i)}_{\nu}(x,\mathcal{R}(\omega))\,. \tag{5.10}$$

As no radiation enters from right

$$i^{(o)}_{\nu}(y,\mathcal{S}(\omega)) = r\,t\,i^{(o)}_{\nu}(x,\mathcal{RS}(\omega)) \tag{5.11}$$

and

$$i^{(o)}_{\nu}(x,\mathcal{RS}(\omega)) = r\,t\,i^{(o)}_{\nu}(y,\mathcal{S}(\omega)) + \frac{|n(y)\cdot\mathcal{R}(\omega)|}{|n(y)\cdot\mathcal{RS}(\omega)|}\left(1-r\right)i^{(i)}_{\nu}(x,\mathcal{R}(\omega))\,. \tag{5.12}$$

Equations (5.11) and (5.12) give

$$i^{(o)}_{\nu}(y,\mathcal{S}(\omega)) = \frac{1}{1-t^2\,r^2}\left(\frac{|n(y)\cdot\mathcal{R}(\omega)|}{|n(y)\cdot\mathcal{RS}(\omega)|}\,r\,t\left(1-r\right)i^{(i)}_{\nu}(x,\mathcal{R}(\omega))\right)\,. \tag{5.13}$$

Substituting (5.13) into Equation (5.10) yields the reflected radiance as a function if the incident radiation and the emission of the plate

$$i_\nu^{(r)}(x,\omega) = r \ \left(1 + \frac{(1-r)^2}{1-t^2\,r^2}t^2\right)\ i_\nu^{(i)}(x,\mathcal{R}(\omega)) \tag{5.14}$$

This equation serves as the definition of the reflectance of the plate

$$r_{\nu,T}^*\ (x,\omega) = r\left(1 + \frac{(1-r)^2}{1-t^2\,r^2}t^2\right)\ . \tag{5.15}$$

The absorptivity of the plate is obtained from energy conservation

$$a_{\nu,T}\ (x,\omega) = 1 - r_{\nu,T}^*\ (x,\omega) - t_{\nu,T}^*\ (x,\omega) = \frac{(1-\ r)(1-t)}{1-rt} \tag{5.16}$$

and the emittance from Kirchhoff's law

$$e_{\nu,T}\ (x,\omega) = a_{\nu,T}\ (x,\omega) \tag{5.17}$$

In the limit $t \to 0$, we obtain the expressions discussed in Chapter 1.

The McMahon solution of the radiation transport equation for plane parallel plates allows us to reformulate the optical properties of a semi-transparent material as boundary conditions for radiation. In this approximation the reflected or transmitted radiance $i_\nu^{(r,t)}(x,\omega)$ is always directed from the plate surface to the outside. Thus in the following the surface normal $n^*(x)$ will always be directed from the solid surface to the outside. As in Equation (1.90) the boundary condition at an arbitrary solid or semi-transparent surface is now

$$i_\nu^{(r,t)}(x,\omega) = \tag{5.18}$$

$$\theta_x^\omega e_{\nu,T}(x,\omega)\ i_{\nu,T}^{BB}(x) + \int_\Omega \frac{|n^*(x)\cdot\omega'|}{|n^*(x)\cdot\omega|}\ \varsigma_{\nu,T}(x,\omega \leftarrow \omega')\ i_\nu^{(i)}(x,\omega')d\omega'$$

The outgoing radiance here has a contribution both from surface emission and from reflection and transmission. The Heaviside function $\theta_x^\omega :=$ $\theta(n^*(x)\cdot\omega)$, which is 1 for positive argument and zero otherwise, expresses that there is only emission in the outside direction with $n^*(x)\cdot\omega > 0$. The specular transmission-reflection function in the McMahon approximation is

$$\varsigma_{\nu,T}^{MM}\ (x,\omega\leftarrow\omega') = \begin{cases} \delta(\omega - \mathcal{S}(\omega')) & n^*(x)\cdot\omega' > \cos\varpi_T \\ \delta(\omega - \mathcal{R}(\omega')) & n^*(x)\cdot\omega' > 0,\ n^*(x)\cdot\omega' < \cos\varpi_T \\ r_{\nu,T}^*\ (x,\omega')\delta(\omega - \mathcal{R}(\omega')), & n^*(x)\cdot\omega' < 0,\ n^*(x)\cdot\omega > 0 \\ t_{\nu,T}^*\ (x,\omega')\delta(\omega - \mathcal{S}(\omega')), & n^*(x)\cdot\omega' < 0,\ n^*(x)\cdot\omega < 0 \end{cases} \tag{5.19}$$

$\varpi_T = \arcsin(n_{\nu,T}/n'_{\nu,T})$ is the angle of total reflection. Later, the diffuse approximation will be introduced. In this approximation, the generalized transmission-reflection function will be

$$\varsigma_{\nu,T}^{DA}(x,\omega \leftarrow \omega') = \begin{cases} \delta(\omega - \omega') & n^*(x) \cdot \omega' > 0 \\ r_{\nu,T}^*\,(x,n^*(x))\frac{|n^*(x) \cdot \omega|}{\pi} & n^*(x) \cdot \omega' < 0, \; n^*(x) \cdot \omega > 0 \\ t_{\nu,T}^*\,(x,n^*(x))\delta(\omega - \omega') & n^*(x) \cdot \omega' < 0, \; n^*(x) \cdot \omega < 0 \end{cases}$$

$$(5.20)$$

In the approximation of surface to surface radiation, the boundary conditions are usually formulated for radiation intensities instead of the radiance. In this case Equation (5.18) becomes

$$I_\nu^{(r,t)}(x,\omega) = \qquad\qquad\qquad\qquad\qquad\qquad\qquad (5.21)$$

$$\theta_x^\omega e_{\nu,T}(x,\omega)\,|n^*(x) \cdot \omega|\, c\, i_{\nu,T}^{BB}(x) + \int_\Omega \varsigma_{\nu,T}(x,\omega \leftarrow \omega')I_\nu^{(i)}(x,\omega')d\omega'$$

A self-consistent equation for the radiation transport can now be obtained by combining the solution of the transport equation (1.75) with the above boundary condition to give

$$I_\nu(x,\omega) = \qquad\qquad\qquad\qquad\qquad\qquad\qquad (5.22)$$

$$\theta_x^\omega e_{\nu,T}(x,\omega)\,|n^*(x) \cdot \omega|\, c\, i_{\nu,T}^{BB}(x) + \int_\Omega \varsigma_{\nu,T}(x,\omega \leftarrow \omega')I_\nu(x - s\omega',\omega')d\omega'$$

where $I_\nu(x,\omega)$ is the radiation intensity of the surface.

5.2.2   Example: Single Coating Layer

We now present as an example the industrially important case of a single coating layer covering a semi-transparent bulk material. The generic situation is shown in Figure 5.2. The media are characterized by the complex refractive indices

$$\tilde{n}_1 = n_1 - ik_1 \quad , \quad \tilde{n}_2 = n_2 - ik_2 \quad , \quad \tilde{n}_3 = n_3 - ik_3 . \qquad (5.23)$$

These are the fundamental quantities used to describe the optical properties of a material. The real part describes the refraction of radiation, while the imaginary part describes the absorption of radiation in the material. In the case of an ideal surface, the reflectivity, emissivity, absorptivity and transmissivity of the surface can be derived from the complex refractive index, including the spectral and directional dependence. An ideal surface is perfectly plane and is composed of a single material without contamination. In the ideal case or in the case of a well defined surface model, one can use the library of complex refractive indices to calculate the effective optical properties of the surfaces.

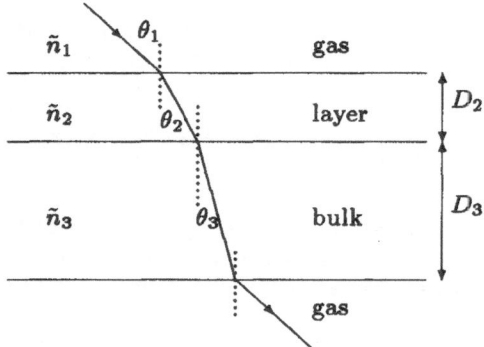

Figure 5.2: Semi-transparent bulk material with surface layer

The ray angles θ_1, θ_2 and θ_3 relative to the surface normal are related by the generalized Snell's law, where the angles become complex. This causes some problems if the imaginary part becomes significant, because the waves in the medium become inhomogeneous, i.e. the surfaces of constant phase and of constant amplitude are different [16]($\S13.2$). Then it is not possible to determine the ray directly from Snell's law. However, to a good approximation, the real part of the angle of refraction may still be used to estimate the ray path. In most cases, either the surface layer is weakly absorbing and the bulk is completely absorbing, or the surface layer is completely absorbing. For cases in which the surface layer and the bulk are both partially absorbing, the ray trajectory will have an error. But the precise knowledge of this trajectory is not of key importance. The generalized Snell's law is

$$\tilde{n}_1 = n_1 \quad , \quad \tilde{n}_2 = n_2 - ik_2 \quad , \quad \tilde{n}_3 = n_3 - ik_3 \tag{5.24}$$

$$n_1 \sin\theta_1 = \tilde{n}_2 \sin\tilde{\theta}_2 \quad , \quad \tilde{n}_2 \sin\tilde{\theta}_2 = \tilde{n}_3 \sin\tilde{\theta}_3$$

The reflectivity and transmissivity of an interface connecting two media may be defined in terms of the reflected and transmitted amplitudes of the $s-$ and $p-$ polarized waves. For an interface connecting medium i with medium j, these amplitudes are given in terms of the complex refractive indices and the complex refractive angles by

$$A_{i,s}^{(r)} = \frac{\cos\tilde{\theta}_i\tilde{n}_i - \cos\tilde{\theta}_j\tilde{n}_j}{\cos\tilde{\theta}_i\tilde{n}_i + \cos\tilde{\theta}_j\tilde{n}_j} \qquad A_{i,p}^{(r)} = \frac{\cos\tilde{\theta}_i\tilde{n}_j - \cos\tilde{\theta}_j\tilde{n}_i}{\cos\tilde{\theta}_i\tilde{n}_j + \cos\tilde{\theta}_j\tilde{n}_i} \tag{5.25}$$

$$A_{i,s}^{(t)} = \frac{2\cos\tilde{\theta}_i\tilde{n}_i}{\cos\tilde{\theta}_i\tilde{n}_i + \cos\tilde{\theta}_j\tilde{n}_j} \qquad A_{i,p}^{(t)} = \frac{2\cos\tilde{\theta}_i\tilde{n}_i}{\cos\tilde{\theta}_i\tilde{n}_j + \cos\tilde{\theta}_j\tilde{n}_i}$$

Let z^* demote the complex conjugate of z. Then the analogue to the Fresnel formula (5.2) in this case is

$$r_{\nu,T} = \frac{1}{2} \left(\mathcal{A}_{1,s}^{(r)} \mathcal{A}_{1,s}^{(r)*} + \mathcal{A}_{1,p}^{(r)} \mathcal{A}_{1,p}^{(r)*} \right) \tag{5.26}$$

Interference in Bulk and Layer

The interference of the multiply reflected light with wavelength λ within one layer leads to phase factors δ_2 and δ_3 in the layer of thickness D_2 and in the bulk material of thickness D_3

$$\delta_2 = 2\pi \frac{D_2}{\lambda} \tilde{n}_2 \cos \tilde{\theta}_2 \qquad \delta_3 = 2\pi \frac{D_3}{\lambda} \tilde{n}_3 \cos \tilde{\theta}_3 \tag{5.27}$$

As discussed in [6] the reflected amplitude is

$$\mathcal{A}^{(r)} = \frac{\mathcal{A}_1^{(r)} + \mathcal{A}_2^{(r)} e^{-2i\delta_2} + \mathcal{A}_3^{(r)} e^{-2i(\delta_2+\delta_3)} + \mathcal{A}_1^{(r)} \mathcal{A}_2^{(r)} \mathcal{A}_3^{(r)} e^{-2i\delta_3}}{1 + \mathcal{A}_1^{(r)} \mathcal{A}_2^{(r)} e^{-2i\delta_2} + \mathcal{A}_1^{(r)} \mathcal{A}_3^{(r)} e^{-2i(\delta_2+\delta_3)} + \mathcal{A}_2^{(r)} \mathcal{A}_3^{(r)} e^{-2i\delta_3}} \tag{5.28}$$

and the transmitted amplitude is

$$\mathcal{A}^{(t)} = \frac{\mathcal{A}_1^{(t)} \mathcal{A}_2^{(t)} \mathcal{A}_3^{(t)} e^{-2i(\delta_2+\delta_3)}}{1 + \mathcal{A}_1^{(r)} \mathcal{A}_2^{(r)} e^{-2i\delta_2} + \mathcal{A}_1^{(r)} \mathcal{A}_3^{(r)} e^{-2i(\delta_2+\delta_3)} + \mathcal{A}_2^{(r)} \mathcal{A}_3^{(r)} e^{-2i\delta_3}} . \tag{5.29}$$

For unpolarized radiation, the reflectance and the transmittance are given as

$$\overset{\approx}{r}{}^*_{\nu,T} = \frac{1}{2} \left(\mathcal{A}_s^{(r)} \mathcal{A}_s^{(r)*} + \mathcal{A}_p^{(r)} \mathcal{A}_p^{(r)*} \right) \quad \overset{\approx}{t}{}^*_{\nu,T} = \frac{1}{2} \left(\mathcal{A}_s^{(t)} \mathcal{A}_s^{(t)*} + \mathcal{A}_p^{(t)} \mathcal{A}_p^{(t)*} \right) \tag{5.30}$$

The method has a straightforward generalization to the case of an arbitrary number of homogeneous layers and is known as the matrix method of Hayfield and White [6].

No Interference in Bulk

In most cases the bulk material will be large compared to the coherence length of the radiation, which is usually on the order of $0.1\ mm$. Therefore there should be no interference effects within the bulk layer. A formula without such interference effects can be obtained from the reflectance and the transmittance of Equation (5.30) by averaging over the phase change, i.e.

$$\tilde{r}^*_{\nu,T} = \frac{1}{2\pi} \int_0^{2\pi} \overset{\approx}{r}{}^*_{\nu,T}\ d(2\delta_3) \qquad \tilde{t}^*_{\nu,T} = \frac{1}{2\pi} \int_0^{2\pi} \overset{\approx}{t}{}^*_{\nu,T}\ d(2\delta_3) \tag{5.31}$$

Squaring (5.28) with the abbreviations

$$\mathcal{A}^{(r)} = \frac{\alpha + \beta e^{-2i\phi_3}}{\rho + \sigma e^{-2i\phi_3}} \tag{5.32}$$

$$\alpha = \mathcal{A}_1^{(r)} + \mathcal{A}_2^{(r)}\tilde{d}_2 \qquad \beta = \mathcal{A}_3^{(r)}\left(\tilde{d}_2 + \mathcal{A}_1^{(r)}\mathcal{A}_2^{(r)}\right)d_3$$

$$\rho = 1 + \mathcal{A}_1^{(r)}\mathcal{A}_2^{(r)}\tilde{d}_2 \qquad \sigma = \mathcal{A}_3^{(r)}\left(\mathcal{A}_1^{(r)}\tilde{d}_2 + \mathcal{A}_2^{(r)}\right)d_3$$

$$\tilde{d}_2 := e^{-2i\delta_2} \qquad d_3 e^{-2i\phi_3} := e^{-2i\delta_3}$$

gives

$$\mathcal{A}^{(r)}\mathcal{A}^{(r)*} = \frac{\alpha\alpha^* + \beta\beta^* + (\alpha\beta^* + \alpha^*\beta)\cos(2\phi_3) + i(\alpha\beta^* - \alpha^*\beta)\sin(2\phi_3)}{\rho\rho^* + \sigma\sigma^* + (\rho\sigma^* + \rho^*\sigma)\cos(2\phi_3) + i(\rho\sigma^* - \rho^*\sigma)\sin(2\phi_3)} . \tag{5.33}$$

For

$$\mathcal{A}^{(t)} = \frac{\epsilon}{\rho + \sigma e^{-2i\phi_3}} \tag{5.34}$$

with

$$\epsilon = \mathcal{A}_1^{(t)}\mathcal{A}_2^{(t)}\mathcal{A}_3^{(t)}e^{-2i(\delta_2+\delta_3)}$$

one obtains

$$\mathcal{A}^{(t)}\mathcal{A}^{(t)*} = \frac{\epsilon\epsilon^*}{\rho\rho^* + \sigma\sigma^* + (\rho\sigma^* + \rho^*\sigma)\cos(2\phi_3) + i(\rho\sigma^* - \rho^*\sigma)\sin(2\phi_3)} \tag{5.35}$$

The integrals given in Equation (5.31) can now be evaluated with the residue theorem, and one obtains the reflectance and transmittance for a coated material with interference effects in the layer and without interference effects in the bulk

$$\tilde{r}^*_{\nu,T} = \frac{|\alpha|^2 + |\beta|^2}{||\rho|^2 - |\sigma|^2|} - \frac{2\,\Re(\alpha\beta^*\rho\sigma^*)}{|\rho|^2\,||\rho|^2 - |\sigma|^2|} \tag{5.36}$$

and

$$\tilde{t}^*_{\nu,T} = \frac{|\epsilon|^2}{||\rho|^2 - |\sigma|^2|} \tag{5.37}$$

Here $\Re(z)$ is the real part of z. Figure 5.8 shows the optical properties of a variety of materials with the effects of surface layers.

No Interference in Bulk and Layer

When the thickness of the surface layer is also large compared to the coherence length, the reflectance and transmittance are described by

$$r^*_{\nu,T} = \frac{1}{2\pi}\int_0^{2\pi} \tilde{r}^*_{\nu,T}\; d(2\delta_2) \qquad t^*_{\nu,T} = \frac{1}{2\pi}\int_0^{2\pi} \tilde{t}^*_{\nu,T}\; d(2\delta_2)\,. \tag{5.38}$$

Equations (5.38) are coupled to Equations (5.30) and (5.31). These formulas account for the multiple reflections at the boundary from medium 1 to medium 2, medium 2 to medium 3 and medium 3 to medium 1 without any interference effect, as well as absorption inside media 1 and 2. Evaluation of the integrals gives

$$r^*_{\nu,T} = R_{12} + \frac{(1 - R_{12})^2 T_2^2 R_{23}}{1 - R_{12}T_2^2 R_{23}} \tag{5.39}$$

$$+\frac{1}{1 - R_{12}T_2^2 R_{23}} \frac{(1 - R_{12})^2 T_2^2 (1 - R_{23})^2 T_3^2 R_{31}}{(1 - R_{12}T_2^2 R_{23})(1 - R_{23}T_3^2 R_{31}) - R_{12}T_2^2(1 - R_{23})^2 T_3^2 R_{31}}$$

$$t^*_{\nu,T} = \frac{(1 - R_{12})T_2(1 - R_{23})T_3(1 - R_{31})}{(1 - R_{12}T_2^2 R_{23})(1 - R_{23}T_3^2 R_{31}) - R_{12}T_2^2(1 - R_{23})^2 T_3^2 R_{31}} \tag{5.40}$$

The reflectivities R_{ij} at the boundaries are calculated from (5.26), while the transmissivities T_i are defined as

$$T_i = \exp\left(-\frac{4\pi k_i D_i}{\lambda \Re(cos(\tilde{\theta}_i))}\right) \tag{5.41}$$

The McMahon formulas (5.9) and (5.15) are a special case of (5.39) and (5.40) when either medium 2 or 3 is a vacuum.

5.3 Optical Properties of Surfaces

A good description of the optical properties of the surfaces in the RTP processing environment is a key issue in carrying out a predictive simulation. Only the Monte Carlo method is able to exploit this information in detail. Properties which can only be treated within a Monte Carlo simulation are the details of the spectral dependence and the details of the angular dependence. Since the necessary optical properties are usually not available completely, it is desirable to formulate as much as possible reasonable models for these properties as functions of some fundamental quantity.

In many cases, the surfaces are not ideal. This can be the case if the bulk material is inhomogeneous, if the bulk has a granular structure such as in amorphous or polycrystalline structures or if the surface is rough. The optical properties of these nonideal surfaces can still be obtained from the complex refractive index under some assumptions. One regime is given by surfaces with a microstructure which is small compared to the wavelength of the radiation. Then the effective medium theory can be applied; this involves averaging the dielectric properties of the participating media (see e.g. [78], [5]). In the case that the surface roughness is not small compared to the wavelength of the radiation, the multiple scattering of radiation invalidates the assumptions of the effective medium theory. In this regime of rough surfaces, several techniques exist to estimate the influence of the structure on the optical property and reflection function [59], [26]. The best known is the Kirchhoff approximation, which performs a coherent superposition of the electromagnetic fields reflected from a surface with a given model geometry. These methods have found their way into the field computer graphics [37] to allow modeling of realistic surfaces.

The spectral dependence of the refractive index of many metals and semiconductors can be found in [78], [77]. There is also software which calculates the refractive index from solid state models [4]. The temperature dependence of the optical properties, which is of particular importance for semiconductors, is often difficult to determine. For silicon, good models for the temperature dependence of the spectral refractive index exist [90].

When tabulating the refractive index in a database, one must keep in mind that the dependence of the refractive index on wavelength is in general not smooth and is best captured by dividing the spectral variable into intervals. It is most convenient to choose these intervals evenly spaced on a logarithmic scale in wave number. The temperature dependence of the refractive index is, however, in general smooth and can be accounted for by a series expansion in every interval. One possible parameterization of the complex refractive index is the following, where i denotes the spectral interval

$$n_{\nu(i),T} = n_i^{(0)} + n_i^{(1)} \left(\frac{T - 300^\circ K}{1000^\circ K} \right) + n_i^{(2)} \left(\frac{T - 300^\circ K}{1000^\circ K} \right)^2 \qquad (5.42)$$

$$k_{\nu(i),T} = \exp \left(k_i^{(0)} \left(\frac{T - 300^\circ K}{1000^\circ K} \right) \right) \left(k_i^{(1)} + k_i^{(2)} \left(\frac{T - 300^\circ K}{1000^\circ K} \right) \right)$$

The parameterization of the absorption coefficient takes the strong temperature dependence, characteristic of semiconductors, into account. Figure 5.5 shows n_i and k_i for silicon and the resulting emissivity in normal direction for several temperatures. The coefficients in this case were fitted to a model from [90]. Figures 5.5 - 5.7 further show the emissivity and reflectivity of a variety of data important for the simulation of RTP processing equipment. The data shown extend over a logarithmic scale from $10^2 cm^{-1}$ to $10^6 cm^{-1}$ (wavelength from 10^{-2} microns to 100 microns) divided into 60 intervals. This allows the treatment of radiation from $250^\circ K$ to $6000^\circ K$.

In the above sections, formulas for the calculation of the optical properties of surfaces for some important cases were presented. Of course it is not possible to calculate a priori the optical properties of a surface to very high precision, as there are usually too many unknowns concerning the surface characteristics. But the requirements on the knowledge of the optical properties for the purpose of modeling radiative heat transfer are less strict than for example for analyzing ellipsometric measurements, since the important quantities in the case of heat transfer are integrals over the radiation distribution. From this point of view, the calculation of the optical properties with models as above should be quite good. It is certainly the best which may be achieved under the constraints of a practical simulation.

5.4 Solution of the Integral Equation

We now address the question of solutions to the radiation transport equation. First we derive a formal solution of the equation with a general transmission-radiation function $\varsigma_{\nu,T}(x, \omega \leftarrow \omega')$ and outline a general strategy for evaluating the solution. A key difficulty here is the implicit dependence of all quantities on temperature. The special cases of specular reflection and diffuse reflection are then considered, as they allow certain simplifications to be made.

5.4.1 The Series Solution

Earlier in this chapter the fundamental transport equation in the McMahon approximation

$$I_\nu(x, \omega) = \tag{5.43}$$

$$\theta_x^\omega e_{\nu,T}(x, \omega) \left| n^*(x) \cdot \omega \right| c \, i_{\nu,T}^{BB}(x) + \int_\Omega \varsigma_{\nu,T}(x, \omega \leftarrow \omega') I_\nu(x - s\omega', \omega') d\omega'$$

was derived (Equation (5.22)), and the optical properties of the surfaces were defined. The next step is to find a method to solve the equation. We derive now a formal series solution based on iteration. The Monte Carlo method used to numerically evaluate this solution is described in the next chapter.

We begin by defining certain functions which arise naturally in the analysis of Equation (5.43) (see Figure 5.3). The parameter s has been defined as the distance along direction $-\omega'$ from the point x to the first intersection with a surface at point y. Thus $y = x - s\omega'$. It is then possible to define the ray tracing function

$$\begin{aligned} X: \quad & D \times \Omega^+ \longrightarrow D \\ & (y, \omega') \longmapsto x = X(y, \omega') \,. \end{aligned} \tag{5.44}$$

X is then the target point as a function of the originating point y and the direction of ray ω'. Another function of interest is the direction of a straight line connecting the originating and target points y and x

$$\begin{aligned} \phi: \quad & D \times D \longrightarrow \Omega \\ & (y, x) \longmapsto \omega' = \phi(y, x) \,. \end{aligned} \tag{5.45}$$

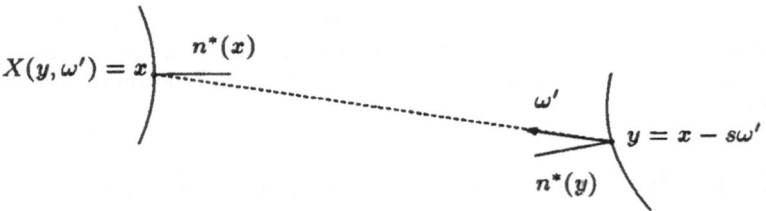

$$X(y, \omega') = x$$

Figure 5.3: Geometry for the coordinate transformation

The coordinate transformation from ω' describing a surface area on the target by cone angle to a coordinate chart on the surface introduces the following Jacobian terms

$$\left|\frac{\partial\phi(y, x)}{\partial x}\right| = \frac{|n^*(x) \cdot \omega'|}{r^2} \quad , \quad \left|\frac{\partial\phi(y, x)}{\partial y}\right| = \frac{|n^*(y) \cdot \omega'|}{r^2} \qquad (5.46)$$

where r is the distance from y to x.

The goal is now to introduce the surface position as an integration variable in Equation (5.43). Let $f(\omega')$ represent the integrand in this equation and let $\mathcal{I}$ be the value of the integral. The first step is to change the integral over ω' to an integral over $y \in D^{vis}(x)$, where $D^{vis}(x)$ is the set of point on the surface visible from the point x without obstruction. This introduces a Jacobian term $|\partial\phi(y, x)/\partial y|$. Thus

$$\mathcal{I} = \int_{D^{vis}(x)} f(\phi(y, x)) \left|\frac{\partial\phi(y, x)}{\partial y}\right| dy \ . \qquad (5.47)$$

The next step is to rewrite the integrand here as a function of $x' \in D^{vis}(y)$ times a delta function $\delta(x - x')$ and integrate over $D^{vis}(y)$:

$$\mathcal{I} = \int_{D^{vis}(x)} \int_{D^{vis}(y)} \delta(x - x') f(\phi(y, x')) \left|\frac{\partial\phi(y, x')}{\partial y}\right| dx' \, dy \ . \qquad (5.48)$$

Finally, the change of variables $x' = X(y, \omega')$ is used so that the interior integral is now integrated over the domain Ω_y with integration variable ω'. This introduces another Jacobian term of the form $|\partial\phi(y, x))/\partial x|^{-1}$ to be evaluated at $x = X(y, \omega')$. Thus

$$\mathcal{I} = \int_{D^{vis}(x)} \int_{\Omega_y} \delta(x - X(y, \omega')) f(\omega') \left|\frac{\partial\phi(y, X)}{\partial y}\right| \left|\frac{\partial\phi(y, X)}{\partial X}\right|^{-1} d\omega' \, dy \ . \qquad (5.49)$$

Since the delta function puts the integrand to zero when the view from x to y is obstructed, the domain of integration for y can be extended from

$D^{vis}(x)$ to D. The delta function also allows us to write x in place of X in the integrand. The result is that Equation (5.43) may now be written

$$I_\nu(x,\omega) = \theta_x^\omega e_{\nu,T}(x,\omega) \, |n^*(x) \cdot \omega| \, c \, i_{\nu,T}^{BB}(x) \tag{5.50}$$
$$+ \int_D \int_{\Omega_y} \varsigma_{\nu,T}(x,\omega \leftarrow \omega') \delta(x - X(y,\omega')) I_\nu(y,\omega') \frac{|n^*(y) \cdot \omega'|}{|n^*(x) \cdot \omega'|} \, d\omega' dy$$

The corresponding equation for the radiance is

$$i_\nu(x,\omega) = \theta_x^\omega e_{\nu,T}(x,\omega) \, i_{\nu,T}^{BB}(x) \tag{5.51}$$
$$+ \int_D \int_{\Omega_y} \varsigma_{\nu,T}(x,\omega \leftarrow \omega') \delta(x - X(y,\omega')) \, i_\nu(y,\omega') \frac{|n^*(y) \cdot \omega'|}{|n^*(x) \cdot \omega|} \, d\omega' dy$$

The solution of the integral equation (5.51) can be obtained by iteration. The solution of (5.50) will differ only by the additional cosine factors. If the series of functions $i_\nu^{(k)}(x,\omega)$ are defined by

$$i_\nu^{(0)}(x,\omega) = \theta_x^\omega e_{\nu,T}(x,\omega) \, i_{\nu,T}^{BB}(x) \tag{5.52}$$

$$i_\nu^{(k)}(x,\omega) =$$
$$\int_D \int_{\Omega_y} \varsigma_{\nu,T}(x,\omega \leftarrow \omega') \delta(x - X(y,\omega')) \, i_\nu^{(k-1)}(y,\omega') \frac{|n^*(y) \cdot \omega'|}{|n^*(x) \cdot \omega|} \, d\omega' dy$$

then it can be shown that

$$i_\nu(x,\omega) = \sum_{k=0}^\infty i_\nu^{(k)}(x,\omega) \tag{5.53}$$

is formally a solution of the equation (5.51). By repeated substitution, $i_\nu^{(k)}(x,\omega)$ can be expressed as

$$i_\nu^{(k)}(x,\omega) = \int_{[\Omega]^k} \int_{[D]^k} \varsigma_{\nu,T}(x,\omega \leftarrow \omega_{k-1}) \cdots \varsigma_{\nu,T}(y_1,\omega_1 \leftarrow \omega_0) \frac{|n^*(y_0) \cdot \omega_0|}{|n^*(x) \cdot \omega|}$$
$$\times \delta(x - X(y_{k-1},\omega_{k-1})) \cdots \delta(y_1 - X(y_0,\omega_0))$$
$$\times \theta_{y_0}^{\omega_0} e_{\nu,T}(y_0,\omega_0) \, i_{\nu,T}^{BB}(y_0) \, d\omega_{k-1} dy_{k-1} \cdots d\omega_0 dy_0 \tag{5.54}$$

$i_\nu^{(k)}(x,\omega)$ can be interpreted as the radiance leaving the surface from point x in the direction ω due to rays which have undergone k reflections.

Generally, the quantity of interest in the modeling of radiative heat transfer is $R(x)$, net radiation flux density through a surface at x. This is the spectrally integrated difference between $A_\nu(x)$, the spectral intensity of the radiation absorbed by the surface and $E_\nu(x)$, the spectral intensity of the radiation emitted by the surface. $A_\nu(x)$ can be expressed as

$$A_\nu(x) = \int_D \int_{\Omega_y} \delta(x - X(y,\omega')) a_{\nu,T}(x,\omega') \, c \, |n^*(x) \cdot \omega'| i_\nu(y,\omega') \, d\omega' \, dy \,. \tag{5.55}$$

This is just the total radiance arriving at x from anywhere else in the reactor (hence the integration over Ω_y and D) weighted by $a_{\nu,T}(x,\omega')$, the percentage of radiance which is absorbed. The factor $c\,|n^*(x)\cdot\omega'|$ converts the radiance to intensity. With the help of Equation (5.53), the intensity of the absorbed radiation may also be expressed as an infinite sum with terms given by

$$A_\nu^{(k)}(x) = \tag{5.56}$$

$$\int_{[\Omega]^{k+1}}\int_{[D]^{k+1}} \theta_x^\omega\, a_{\nu,T}(x,\omega_k)\delta(x - X(y_k,\omega_k))\varsigma_{\nu,T}(y_k,\omega_k \leftarrow \omega_{k-1})$$
$$\times\delta(y_k - X(y_{k-1},\omega_{k-1}))\cdots\varsigma_{\nu,T}(y_1,\omega_1 \leftarrow \omega_0)\delta(y_1 - X(y_0,\omega_0))$$
$$\times\theta_{y_0}^{\omega_0}\, e_{\nu,T}(y_0,\omega_0)\,|n^*(y_0)\cdot\omega_0|\,c\,i_{\nu,T}^{BB}(y_0)\,d\omega_k dy_k \cdots d\omega_0 dy_0$$
$$=: \int_D K_{\nu,T}^{(k)}(x,y_0)\frac{c}{\pi}i_{\nu,T}^{BB}(y_0)\,dy_0$$

The integration over the spectrum yields the absorbed intensity $A^{(k)}(x)$

$$\begin{aligned}
A^{(k)}(x) &= \int_0^\infty \int_D K_{\nu,T}^{(k)}(x,y_0)\frac{c}{\pi}i_{\nu,T}^{BB}(y_0)dy_0 d\nu \tag{5.57}\\
&= \int_0^\infty \int_D \frac{K_{\nu,T}^{(k)}(x,y_0)\frac{c}{\pi}i_{\nu,T}^{BB}(y_0)}{\sigma T^4(y_0)}\,\sigma T^4(y_0)dy_0 d\nu\\
&=: \int_D \tilde{K}_T^{(k)}(x,y_0)\sigma T^4(y_0)\,dy_0
\end{aligned}$$

In order to relate this to the total absorbed intensity at x we define

$$\tilde{K}_T(x,y_0) = \sum_{k=0}^\infty \tilde{K}_T^{(k)}(x,y_0) \tag{5.58}$$

where $\tilde{K}_T^{(k)}(x,y_0)$ is explicitly defined as

$$\tilde{K}_T^{(k)}(x,y_0) = \tag{5.59}$$

$$\int_0^\infty \int_{[\Omega]^{k+1}}\int_{[D]^k} \theta_x^\omega\, a_{\nu,T}(x,\omega_k)\delta(x - X(y_k,\omega_k))\varsigma_{\nu,T}(y_k,\omega_k \leftarrow \omega_{k-1})$$
$$\times\delta(y_k - X(y_{k-1},\omega_{k-1}))\cdots\varsigma_{\nu,T}(y_1,\omega_1 \leftarrow \omega_0)\delta(y_1 - X(y_0,\omega_0))$$
$$\times\theta_{y_0}^{\omega_0}\, e_{\nu,T}(y_0,\omega_0)\,|n^*(y_0)\cdot\omega_0|\,\frac{c\,i_{\nu,T}^{BB}(y_0)}{\sigma T^4(y_0)}d\omega_k dy_k \cdots d\omega_0 d\nu$$

By Equation (1.73) the emitted intensity $E(x)$ is

$$\begin{aligned}
E(x) &= \int_0^\infty \int_\Omega \theta_x^\omega e_{\nu,T}(x,\omega)\,|n^*(n)\cdot\omega|\,c\,i_{\nu,T}^{BB}(x)d\omega d\nu \tag{5.60}\\
&= e_T^{eff}(x)\sigma T^4(x) = \int_D \delta(x - y_0)e_T^{eff}(y_0)\sigma T^4(y_0)dy_0
\end{aligned}$$

where $e_T^{eff}(x)$ is the hemispherically and spectrally averaged emissivity. The net radiation flux density $R(x) = A(x) - E(x)$ through a surface at point x is then

$$R(x) = \int_D \Phi_T(x, y_0) \sigma T^4(y_0) \, dy_0 \qquad (5.61)$$

where

$$\Phi_T(x, y_0) = \left(\tilde{K}_T(x, y_0) - \delta(x - y_0) e_T^{eff}(y_0) \right) \qquad (5.62)$$

is the radiation exchange kernel and represents the solution of the surface to surface heat transfer problem with semi-transparent media in the McMahon approximation. Methods for the explicit calculation of Φ will be discussed in the next chapter.

We now outline several properties of a general simulation strategy. First, for all practical purposes, the surface D must be discretized into disjoint surface elements D_i such that $D = \bigcup_{i=1,N} D_i$. The radiation exchange kernel then becomes a matrix defined as

$$\Phi_T(i, j) = \frac{1}{Area(D_i)} \int_{D_i} \int_{D_j} \Phi_T(x, y) dx dy \qquad (5.63)$$

and the discrete version of Equation (5.61) becomes

$$R(i) = \sum_{j=1,N} \left(\tilde{K}_T(i, j) - \delta_{ij} e_T^{eff}(j) \right) \sigma T^4(j) = \sum_{j=1,N} \Phi_T(i, j) \sigma T^4(j) \, .$$
$$(5.64)$$

A second, more difficult issue is how to treat the temperature dependence of Φ_T. This has been defined, through $\tilde{K}_T^{(k)}(x, y_0)$ and $e_T^{eff}(x)$, to separate out the temperature dependencies of the optical properties from the temperature dependence of the emitted intensity. In case of grey surfaces (i.e. spectrally independent optical properties), Φ_T is independent of temperature. For a fixed geometry, it could be calculated once and for all with a reference temperature distribution T_{ref}. The only temperature dependence of the absorbed intensity would then be the dependence of the emitted intensity on temperature, which is $\sigma T^4(y_0)$.

In the case of real surfaces, however, Φ_T is temperature dependent, and taking this into account explicitly is very important, especially in a transient calculation. This temperature dependence arises from two distinct effects. The first involves the spectral dependency of the transmissivity, reflectivity and absorptivity. Depending on the temperature, a surface will emit radiation with different spectral distributions. The values of the transmissivity, reflectivity and absorptivity then change as a function of the impinging spectral distribution, or equivalently the radiation temperature. An example of this is the absorptivity and reflectivity of radiation emitted from a tungsten-halogen lamp at different radiation temperatures at a steel wall with fixed temperature (see Figure 5.6(c) and the discussion in Section 7.3).

Another example is the transmissivity of radiation at different temperatures through a quartz window with fixed temperature (see Figure 5.7)(b). Thus, even when a surface has a fixed temperature, its optical properties may change due to the nature of the impinging spectral distribution (radiation temperature), which of course depends on the temperature elsewhere in the reactor.

The second effect determining the temperature dependence of Φ_T is due to the fact that optical properties of a surface change with its surface temperature and thus influence the propagation of the radiation. An example is the transmissivity of silicon at $300°K$, which is high for radiation below an energy of $1.12eV \simeq 1.10\mu m$, and silicon at $1000°K$, which has almost zero transmissivity.

It is relatively simple to compute the dependence of Φ_T on the temperature of the emitting surface (i.e., the temperature dependence of the first kind). For a given surface discretization, Φ_T has the form of a matrix

$$\Phi_T = \begin{pmatrix} \Phi_{T_1,T}(1,1) & \Phi_{T_2,T}(1,2) & \cdots & \Phi_{T_N,T}(1,N) \\ \Phi_{T_1,T}(2,1) & \Phi_{T_2,T}(2,2) & \cdots & \Phi_{T_N,T}(2,N) \\ \vdots & \vdots & \ddots & \vdots \\ \Phi_{T_1,T}(N,1) & \Phi_{T_2,T}(N,2) & \cdots & \Phi_{T_N,T}(N,N) \end{pmatrix} \quad (5.65)$$

The dependence on the temperature of the emitting surface j is contained in the matrix elements of column j only. This is denoted by the subscript T_j. T denotes the dependence on the surface temperatures for the transport. In other words, each matrix element depends only on a single radiation temperature. For a linear approximation of the spectral distribution dependence, Φ_T has to be calculated twice for two sets of spectral distributions according to surface temperatures $\left\{T_1^{(0)}, T_2^{(0)}, ..., T_N^{(0)}\right\}$ and $\left\{T_1^{(1)}, T_2^{(1)}, ..., T_N^{(1)}\right\}$ and with fixed optical properties. The selection of surface temperatures should be appropriate for the problem, e.g. $T_j^{(0)}$ and $T_j^{(1)}$ could be the minimal and maximal reasonable values for a reference temperature T_j^{ref}. The Φ_T with explicit dependence on the temperature of the emitting surface is then constructed elementwise in the linear approximation

$$\Phi_{T_j,T}^{lin}(i,j) = t\,\Phi_{T_j^{(1)},T}(i,j) + (1-t)\,\Phi_{T_j^{(0)},T}(i,j) \quad \text{with} \quad t = \frac{T_j - T_j^{(0)}}{T_j^{(1)} - T_j^{(0)}}$$
$$(5.66)$$

The generalization to higher approximations is immediate.

The dependence of Φ_T on the optical properties varying with temperature is not so simple to capture explicitly, since every matrix element in the surface discretized form depends in general on the optical properties and hence temperatures of all surfaces. In cases when the temperature dependence on the transport can be captured by the temperature of a single surface (a quartz window or a mean wafer temperature), a procedure similar to what was just described is conceivable. Otherwise, an iteration procedure must be performed as follows:

- In the first step, a reference temperature distribution $T_{ref}(y_0)$ is estimated. This should preferably be as close as possible to the true temperature distribution of the surfaces. Then $\Phi_{T_{ref}}$ is calculated with a Monte Carlo calculation. This may be improved by calculating $\Phi_{T_{ref}}^{lin}$ as described above.

- The reactor simulation in the CVD simulator is done with the net radiative heat flux source

$$R(x) = \int_D \Phi_{T_{ref}}(x, y_0) \sigma T_{sim}^4(y_0)\, dy_0 \ . \tag{5.67}$$

This reactor simulation includes the conductive and convective heat fluxes. As a results, $T_{sim}(y_0)$ is obtained as the actual surface temperature distribution.

- $T_{ref}(y_0) = T_{sim}(y_0)$ and the calculation of $\Phi_{T_{ref}}$ with Monte Carlo is repeated if the initial and the actual temperature differ too much.

- Eventually a reactor simulation with a new $\Phi_{T_{ref}}$ is done.

In most stationary simulations, the calculation of a single $\Phi_{T_{ref}}$ should be sufficient.

5.4.2 Specular Reflection

In Section 5.2.1 the transmission-reflection function for specularly reflecting surfaces in the McMahon approximation was derived. This is given in Equation (5.19). For semi-transparent materials ($t^* > 0$), this expression stipulates that a certain percentage of the incoming radiation be reflected and a certain percentage be transmitted. This presents a problem for ray tracing algorithms, however, because this corresponds to ray splitting. Thus for each incoming ray, two outgoing rays are created. To avoid exponential growth in the number of rays being tracked, an alternate description of $\varsigma_{\nu,T}^{MM}$ is frequently employed. In this probabilistic formulation, an incoming ray is reflected with probability r^*, transmitted with probability t^*, and absorbed with probability $1 - r^* - t^*$. Mathematically this may be expressed by introducing a random variable z which is uniformly distributed on $[0, 1]$. The transmission-reflection function is then written as

$$\varsigma_{\nu,T}^{MM}(x, \omega \leftarrow \omega') = \int_0^1 \varsigma_{\nu,T}'(z, x, \omega \leftarrow \omega')\, dz \tag{5.68}$$

where the new function $\varsigma'_{\nu,T}$ is defined as

$$\begin{aligned}
\varsigma'_{\nu,T}(z, x, \omega \leftarrow \omega') &= \theta(n^*(x) \cdot \omega')[\theta(\cos \varpi_T - n^*(x) \cdot \omega')\delta(\omega - \mathcal{R}(\omega')) \\
&+\theta(n^*(x) \cdot \omega' - \cos \varpi_T)\delta(\omega - \mathcal{S}(\omega'))] \\
&+\theta(-n^*(x) \cdot \omega')[\theta(r^*_{\nu,T}(x, \omega') - z)\delta(\omega - \mathcal{R}(\omega')) \\
&+\theta(z - r^*_{\nu,T}(x, \omega'))\theta(r^*_{\nu,T}(x, \omega') + t^*_{\nu,T}(x, \omega') - z)\delta(\omega - \mathcal{S}(\omega'))]
\end{aligned} \tag{5.69}$$

Here $\theta(\cdot)$ is the Heavyside function, and Equation (5.68) is a consequence of the fact that

$$r = \int_0^1 \theta(r - z)dz \ . \tag{5.70}$$

The Heavyside functions determine the occurrence of reflection, transmission and absorption according to their respective probabilities and the relation $r^*_{\nu,T} + t^*_{\nu,T} + a^*_{\nu,T} = 1$. The first two terms of Equation (5.69) give the ray path for the case of incidence from inside; this does not depend on the variable z. The last two terms give the reflection and transmission depending on the value of the random number z. If $z > r^*_{\nu,T} + t^*_{\nu,T}$, these terms become zero. The ray is not propagated further, but absorbed. For consistency, this requires that $a_{\nu,T}(x, \omega)$ in Equation (5.59) be replaced by

$$a_{\nu,T}(x, \omega) = \int_0^1 \theta(z - r^*_{\nu,T}(x, \omega) - t^*_{\nu,T}(x, \omega)) \, dz \tag{5.71}$$

It should be remarked that a computational penalty must be paid when this probabilistic approach is used to replace ray splitting. The dimension of the integrals increases by one at each scattering event. In general, the amount of work required to evaluate an integral to a fixed accuracy increases with the dimension of the integral (the curse of dimensionality). It is often the case, however, that the advantages of the probabilistic method outweigh this disadvantage.

A considerable simplification of Equation (5.59) is afforded in the case of specular reflection by the presence of the delta functions in (5.69). Up to the random choice of reflection or transmission, the scattering angle of a ray is fixed by the incoming angle. Thus the entire path of the ray is determined solely by the initial starting point y_0, the initial direction ω_0 and the sequence of random numbers z_k used to determine reflection, transmission or absorption. This means that $\widetilde{K}_T^{(k)}(x, y_0)$ only involves one angular integration over the initial direction. Of course, the introduction of $\varsigma'_{\nu,T}$ requires an additional integration over the probability space I^k (k-dimensional unit cube). The result is that Equation (5.59) becomes

$$\widetilde{K}_T^{(k)}(x, y_0) \;=\; \int_0^\infty \int_{I^k} \int_\Omega \theta_x^\omega \, a_{\nu,T}(x, \omega_k)\delta(x - X(X^{(k)}, \omega_k)) \tag{5.72}$$

$$\times \varsigma_{\nu,T}^{MM}(z_k, X^{(k)}, \omega_k \leftarrow \omega_{k-1}) \cdots \varsigma_{\nu,T}^{MM}(z_1, X^{(1)}, \omega_1 \leftarrow \omega_0)$$
$$\times \theta_{y_0}^{\omega_0} e_{\nu,T}(y_0, \omega_0) |n^*(y_0) \cdot \omega_0| c\, i_{\nu,T}^{BB}(y_0) / \sigma T^4(y_0) d\omega_0 d\nu d^k z$$

5.4.3   Diffuse Reflection

The problem of ray splitting at semi-transparent surfaces also occurs in the case of diffuse transmission-reflection. Again, a probabilistic form of Equation (5.20) may be introduced

$$\varsigma_{\nu,T}^{DA}(x, \omega \leftarrow \omega') = \int_0^1 \varsigma_{\nu,T}''(z, x, \omega \leftarrow \omega')\, dz \tag{5.73}$$

where the new function $\varsigma_{\nu,T}''$ is defined as

$$\varsigma_{\nu,T}''(z, x, \omega \leftarrow \omega') = \tag{5.74}$$
$$\theta(n^*(x) \cdot \omega')\delta(\omega - \omega') + \theta(-n^*(x) \cdot \omega')\theta(r_{\nu,T}^*(x, \omega') - z)|n^*(x) \cdot \omega|$$
$$+\theta(-n^*(x) \cdot \omega')\theta(z - r_{\nu,T}^*(x, \omega'))\theta(t_{\nu,T}^*(x, \omega') + r_{\nu,T}^*(x, \omega') - z)\delta(\omega - \omega')$$

In the diffuse case, however, the angular integrations of Equation (5.59) must be retained. In this case $\widetilde{K}_T^{(k)}(x, y_0)$ is given by

$$\widetilde{K}_T^{(k)}(x, y_0) = \int_0^\infty \int_{I^k} \int_{[\Omega]^{k+1}} \theta_x^\omega\, a_{\nu,T}(x, \omega_k)\delta(x - X(X^{(k)}, \omega_k)) \tag{5.75}$$
$$\times \varsigma_{\nu,T}''(z_k, X^{(k)}, \omega_k \leftarrow \omega_{k-1}) \cdots \varsigma_{\nu,T}''(z_1, X^{(1)}, \omega_1 \leftarrow \omega_0)$$
$$\times \theta_{y_0}^{\omega_0} e_{\nu,T}(y_0, \omega_0) |n^*(y_0) \cdot \omega_0| \frac{c\, i_{\nu,T}^{BB}(y_0)}{\sigma T^4(y_0)} d\omega_k \cdots d\omega_0 d\nu d^k z$$

Just as in the specular reflection case, a Monte Carlo ray tracing algorithm may be use to evaluate $\Phi_T(x, y_0)$ in the diffuse case (using the above $\widetilde{K}_T^{(k)}(x, y_0)$). However, there is another non-Monte Carlo approach for diffuse surfaces that is often computationally more efficient. We outline this procedure now.

We return to Equation (5.22). For radiative heat transfer problems, the key quantity of interest is the total radiation intensity (at a given frequency) arriving at a point. In the final analysis, the direction from which it came is not important. Thus one often works with the the spectral radiosity $B_\nu(x)$, which is defined as the hemispherically integrated radiation intensity

$$B_\nu(x) := \int_{\Omega+} I_\nu(y, \omega) d\omega . \tag{5.76}$$

Here y satisfies $x = X(y, \omega)$.

The spectral radiosity $B_\nu(x)$ is the intensity incident on the surface, while $I_\nu(x, \omega)$ is the intensity leaving the surface. Therefore Equation (5.22) must be modified before it can be used to obtain an equation for the spectral radiosity. We defer the derivation of this new equation for incident intensity until the next section. For now we will present the results of integrating the new equation, which leads to an integral equation for $B_\nu(x)$.

The key feature of diffusely reflecting surfaces, and more generally, surfaces which have a generalized transmission-reflection function according to (5.20), is that the reflected intensity is independent of the angle from which the radiation arrives. This allows the three point reflectance function in Equation (5.89) to come out from under the integral, so that the integral remaining is exactly the spectral radiosity. Under the further assumption that the reflectivity and emissivity of the surface are directionally independent, this equation may be integrated in x to give

$$B_\nu(y) = \int_D f(y, x)\Big(e_{\nu,T}^{hem}(x)\ c\, i_{\nu,T}^{BB}(x) + r_{\nu,T}^{*hem}(x)B_\nu(x)\Big) dx \ . \qquad (5.77)$$

Here

$$f(y, x) = \theta_y^\omega\, \widetilde{t}_{\nu,T}(y, x)\frac{|n^*(y) \cdot \omega||n^*(x) \cdot \omega|}{\pi\ r^2} \qquad (5.78)$$

is a purely geometrically dependent quantity if all surfaces are opaque. Then, $\widetilde{t}_{\nu,T}(y, x)$ is either 1 or 0. When semi-transparent surfaces present, $\widetilde{t}_{\nu,T}(y, x)$ depends on the optical properties of the obstructing semi-transparent surfaces.

Usually, the spectral integration is performed over L spectral intervals or bands $\{[\nu_0, \nu_1], [\nu_1, \nu_2], \cdots, [\nu_{L-1}, \nu_L]\}$. Within each band, all parameters are assumed constant with respect to frequency. Thus Equation (5.77) may be integrated over each band to give a set of L independent equation for radiosity in the individual spectral bands

$$B_l(y) = \int_D f_{l,T}(y, x)\Big(e_{l,T}^{hem}(x)I_{l,T}^{BB}(x) + r_{l,T}^{*hem}(x)B_l(x)\Big) dx \ , \quad l = 1, ..., L \qquad (5.79)$$

When the problem is studied with a discretization of the surface $D = \bigcup_{i=1}^N D_i$, (5.79) is transformed into the matrix equation

$$B_l(i) = \sum_{j=1}^N F_{l,T}(i, j)\Big(e_{l,T}^{hem}(j)I_{l,T}^{BB}(j) + r_{l,T}^{*hem}(j)B_l(j)\Big) \qquad (5.80)$$

where $B_l(i)$ is now the radiation intensity impinging in the spectral band l in units $[W/m^2]$ on surface element i and $I_{l,T}^{BB}(j)$ is the black body intensity leaving the surface element j. The term

$$F_{l,T}(i, j) = \frac{1}{Area(D_j)} \int_{D_j}\int_{D_i} \theta_y^\omega\, \widetilde{t}_{l,T}(y, x)\frac{|n^*(y) \cdot \omega||n^*(x) \cdot \omega|}{\pi\ r^2} dydx \qquad (5.81)$$

is the generalized 'viewfactor' in the presence of semi-transparent surfaces. For a given temperature distribution, the view factor matrix may be computed. The solution to Equation (5.80) may then be obtained by iteration [47], [63]

$$B_l(i) =$$ (5.82)
$$\sum_{j=1}^{N} \left(F_{l,T}(i,j) + \sum_{k=1}^{N} F_{l,T}(i,k) r_{l,T}^{*hem}(k) F_{l,T}(k,j) + \dots \right) e_{l,T}^{hem}(j) I_{l,T}^{BB}(j).$$

The solution may also be expressed explicitly through matrix inversion. If $\vec{B}_l$ is the vector of B's, and $\vec{S}_{l,T}$ is the vector with j^{th} element $e_{l,T}^{hem}(j) I_{l,T}^{BB}(j)$, then $\vec{B}_l$ is given by

$$\vec{B}_l = G_{l,T} \, \vec{S}_{l,T}$$ (5.83)

where the matrix $G_{l,T}$ is defined by

$$G_{l,T} = \left(I - F_{l,T} \, R_{l,T} \right)^{-1} F_{l,T} \, .$$ (5.84)

Here $R_{l,T}$ is a diagonal matrix with $R_{l,T}(j,j) = r_{l,T}^{*hem}(j)$, and I is the $N \times N$ identity matrix.

The diffuse approximation solution of the radiative transport equation can also be used to describe free molecular flow of molecules in a geometry with diffusely reflecting walls and diffuse entrance of the molecules.

$$B_l(i) = \sum_{j=1}^{N} F(i,j) \Big(S_l(j) + r_l(j) B_l(j) \Big)$$ (5.85)

Here the index l labels the different molecular species, $S_l(j)$ is the flux density of the molecules through the entrances in units $\left[molecules/m^2 s \right]$ and $r_l(j)$ is the reflection probability of species l in surface element j. $r_l(j) < 1$ allows for species absorption at the wall.

5.5 The Rendering Equation

There is another formulation of the radiation transport equation which has become popular in the field of computer graphics under the name 'rendering equation' [38]. The starting point in obtaining this equation is the relation (5.18). This is used as a boundary condition for the transport equation (1.75), not to obtain an equation for the radiance leaving a surface as in (1.92), but for the radiance incident on a surface. One further difference between the rendering equation and Equation (1.92) is that for the rendering equation, the target point y need not be the first surface point hit by a ray starting from x in direction ω, but may be on any surface intersected by the ray starting at x. The equation in the McMahon approximation is

$$i_\nu^{(i)}(y,\omega) = \theta_y^\omega \, \tilde{i}(y,x) \tag{5.86}$$

$$\times \left(e_{\nu,T}(x,\omega) \, i_{\nu,T}^{BB}(x) + \int_\Omega \frac{|n^*(x) \cdot \omega'|}{|n^*(x) \cdot \omega|} \varsigma_{\nu,T}(x,\omega \leftarrow \omega') \, i_\nu^{(i)}(x,\omega')d\omega' \right)$$

Since the ray may now be transmitted through a number (say n) of semi-transparent surfaces on the way from x to y, $\tilde{t}(y,x)$ is the product of the transmittances of the intermediate plane parallel surfaces encountered

$$\tilde{t}_{\nu,T}(y,x) = t_{\nu,T}^*(1)t_{\nu,T}^*(2)\cdots t_{\nu,T}^*(n) \tag{5.87}$$

The rendering equation is usually written in terms of surface position instead of direction vectors ω using the change of coordinates given in Equation (5.45) and the corresponding Jacobian terms of (5.46). Moreover, the equation is also generally given in terms of the incident intensity arriving at point y from point x. This is defined in terms of the incident radiance of Equation (5.86) by

$$I_\nu^i(y,x) = c \, i_\nu^{(i)}(y,\omega) \, |n^*(y) \cdot \omega| \, \frac{|n^*(x) \cdot \omega|}{r^2} \tag{5.88}$$

The equation for $I_\nu^i(y,x)$ obtained from (5.86) by multiplying by $|n^*(y) \cdot \omega|$ is then

$$I_\nu^i(y,x) = g(y,x) \left(e_v(y,x) + \int_D \varsigma_{\nu,T}(y,x,z) \, I_\nu^i(x,z)dz \right) \tag{5.89}$$

Here $g(y,x)$ is a geometric term

$$g(y,x) = \theta_y^\omega \frac{\tilde{t}_{\nu,T}(y,x)}{r^2} \,, \tag{5.90}$$

and $e_v(y,x)$ is the unoccluded two point emittance

$$e_v(y,x) = e_{\nu,T}(x,\omega) \, c \, i_{\nu,T}^{BB}(x) \, |n^*(y) \cdot \omega| \, |n^*(x) \cdot \omega| \tag{5.91}$$

The unoccluded three point reflectance function $\varsigma_{\nu,T}(y,x,z)$ is defined in terms of the reflectance-transmittance function as

$$\varsigma_{\nu,T}(y,x,z) = \varsigma_{\nu,T}(x,\omega \leftarrow \omega') \, |n^*(y) \cdot \omega| \tag{5.92}$$

Note that the spectral radiosity defined in Equation (5.76) may now be expressed in terms of $I_\nu^i(y,x)$ as

$$B_\nu(y) = \int_D I_\nu^i(y,x) \, dx \,. \tag{5.93}$$

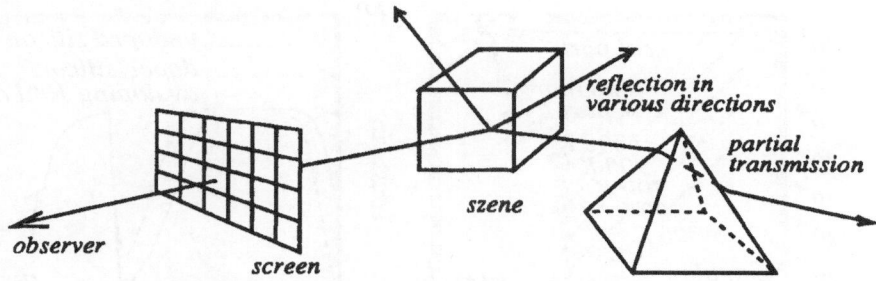

Figure 5.4: View of the observer through the pixel of a screen to the scene and the light sources.

Equation (5.89) is the starting point of many different methods for rendering in computer graphics [38], [53]. Rendering is the calculation of the spectral radiation intensity impinging in a single point y (the viewpoint) which originates from a light source and is transported directly and indirectly through interaction with surfaces (the scene). In other words, it is the calculation of $I_\nu^i(y, \omega)$, where ω points to a pixel of the screen (see Figure (5.4)). The spectral distribution gives the color.

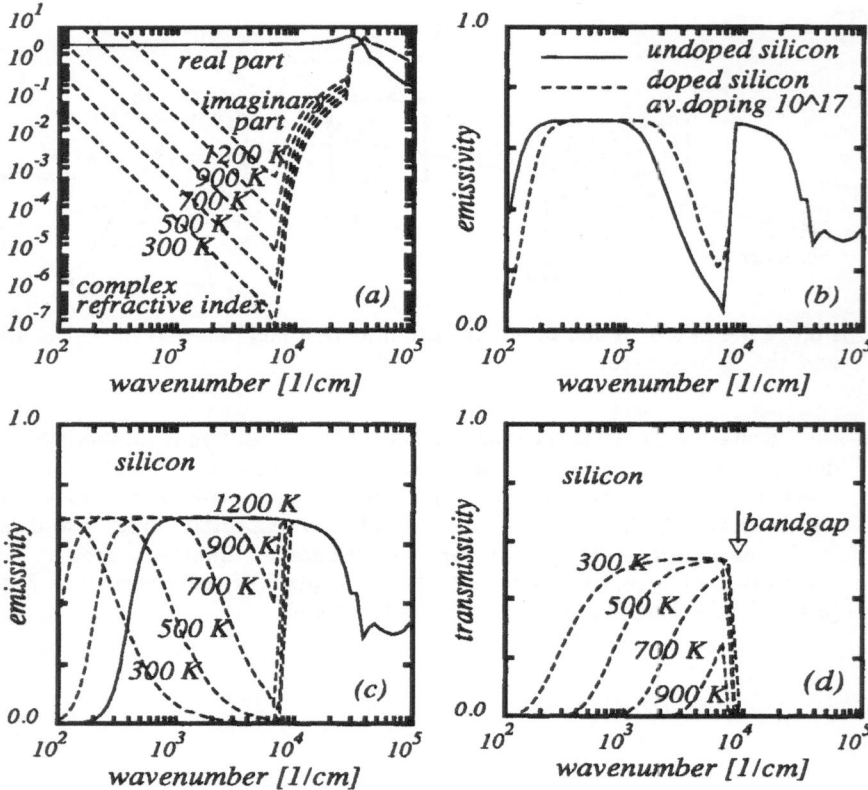

Figure 5.5: Plot (a) shows the temperature dependence of the real and the imaginary part of the refractive index of undoped silicon in the model of [90]. Plot (b) shows the dependence on the doping in this model. The indicated doping is the total amount of dopant averaged over the volume of the wafer. The emissivity at $700°K$ is plotted. Plots (c) and (d) show the emissivity and transmissivity at several temperatures resulting from this model. The model is experimentally validated close to the band gap and loses its predictability in the far infrared.

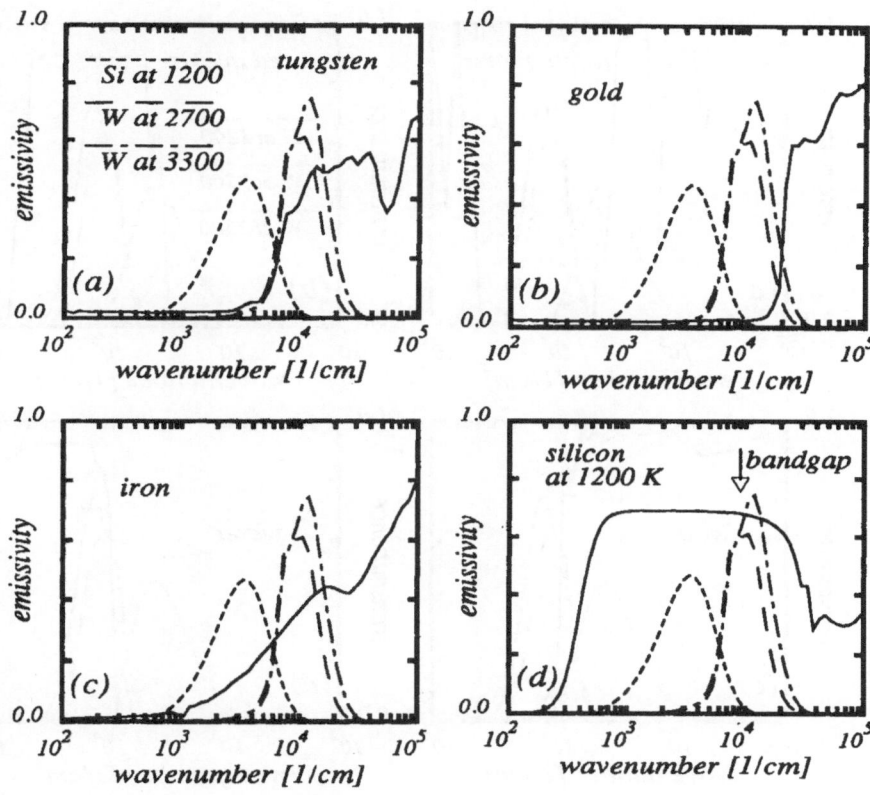

Figure 5.6: This figure shows the emissivity for various opaque material resulting from data of [78] and [77] for the complex refractive index. (a) shows the emissivity of tungsten and the emission spectra from the tungsten-halogen lamp at different filament temperatures. The effective emissivity depends significantly on the filament temperature. The reflectivity of gold, however, depends rather weakly on the filament temperature for achievable values as can be seen from the plot of the emissivity in (b). For iron, the reflectivity depends again on the filament temperature, as can be concluded from (c). The reflectivity of metals is always very high for infrared radiation like the radiation from the moderately heated silicon wafer. The absorptivity of silicon, finally, is rather insensitive on the filament temperature, as can be concluded from (d), provided the silicon itself is heated above $1000°K$.

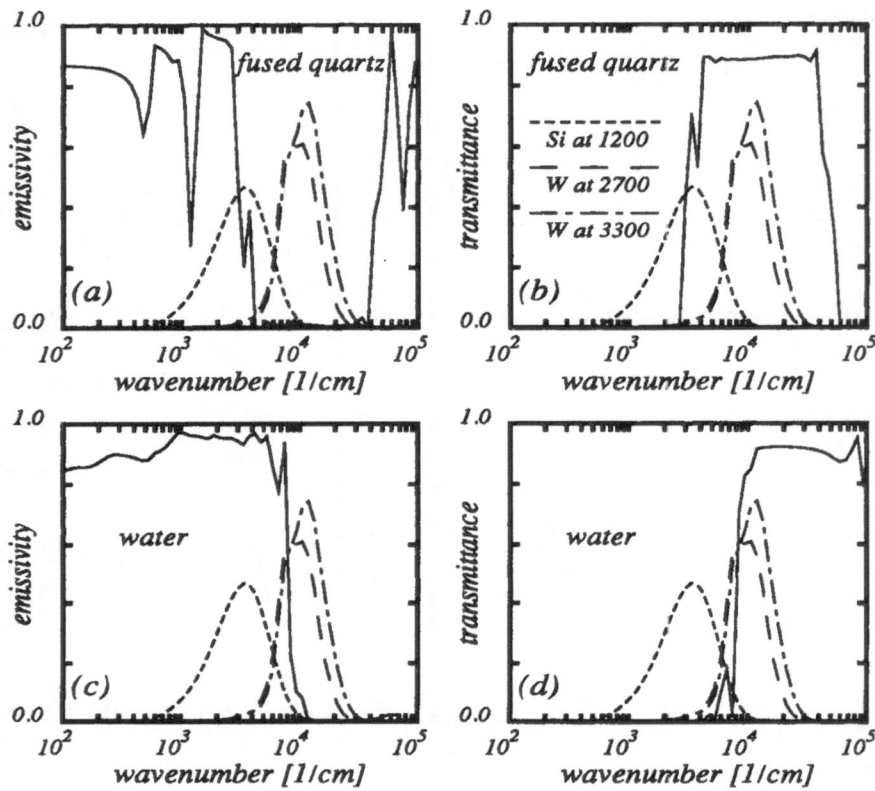

Figure 5.7: This figure shows the emissivity and transmissivity of fused quartz in relation to the emission spectra of the tungsten filament and the heated silicon wafer. (a) and (b) show that fused quartz provides an optical window for the hot radiation from a tungsten filament, but absorbs an appreciable part of the radiation from a heated silicon wafer. The total absorbed amount of radiation depends on details of the absorption spectrum and differs among the different kinds of quartz. Water has a narrower window, as is shown in (c) and (d), and can act as a filter which absorbs almost all the radiation below the bandgap of silicon. The silicon absorptivity of such a radiation is almost temperature independent.

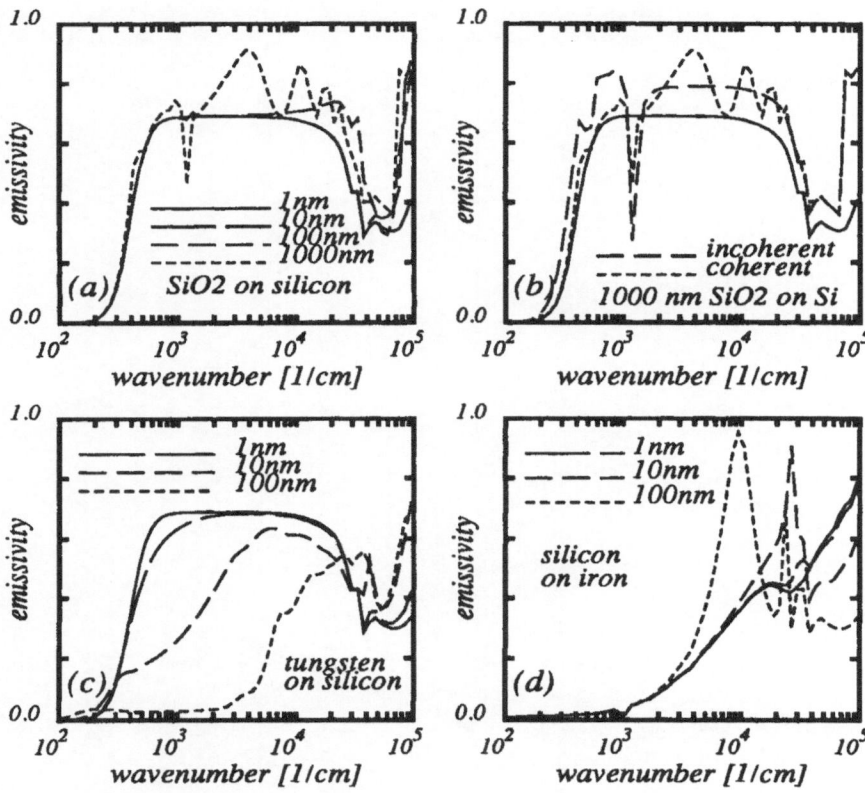

Figure 5.8: This figure shows the emissivity of several coated materials according to the coating models of Section 5.2. Plot (a) shows typical interference effects of SiO_2 on Si when the wavelength is on the order of the layer thickness. Plot (b) shows that the model for incoherent interference approximates the phase average of the model for coherent interference. Plot (c) shows the effect of a tungsten layer, which is semi-transparent when the layer is thin enough, on silicon. The emissivity changes from the emissivity of silicon to the emissivity of tungsten when the layer reaches a few 100 nm. The effect of silicon deposition on a cold steel wall can be seen in plot (d).

Chapter 6

Monte Carlo for Radiation Transport

Here we present a general description of the Monte Carlo approach for solving the radiation transport equations discussed in the previous chapter. A digression is then made into the theory of quasi-Monte Carlo, which is a less widely known technique that provides a significant improvement over the standard Monte Carlo approach in problems like radiation transport. We conclude the chapter with a discussion of how ray tracing is implemented when the radiative heat transfer solver is coupled to a fluid flow simulator.

6.1 Monte Carlo Solution Procedure

We now consider how to numerically evaluate the solution to the radiation transport equation given by Equation (5.61) in the last chapter. To summarize, the net radiation flux density $R(x)$ at a point x on the surface D may be expressed as

$$R(x) = \int_D \Phi_T(x, y)\sigma T^4(y)\, dy \ . \tag{6.1}$$

$\Phi_T(x, y)$ represents the solution of the surface to surface heat transfer problem with semi-transparent media in the McMahon approximation and satisfies the relationship

$$\Phi_T(x, y) = \widetilde{K}_T(x, y) - \delta(x - y)e_T^{eff}(y) \ . \tag{6.2}$$

The quantity $e_T^{eff}(y)$ is the averaged emissivity at a point y on the surface, while $\widetilde{K}_T(x, y)$ is obtained from the iterative solution of the original radiation transport integral equation, and thus is given by the infinite series

$$\widetilde{K}_T(x,y) = \sum_{k=0}^{\infty} \widetilde{K}_T^{(k)}(x,y) \tag{6.3}$$

where

$$\widetilde{K}_T^{(k)}(x,y) = \int_0^{\infty} \int_{[\Omega]^{k+1}} \int_{[D]^k} B^{(k)}(x,y,y_1,\ldots,y_k,\omega_0,\omega_1,\ldots,\omega_k,\nu)$$
$$P(y,\omega_0,\nu)\,dy_k\ldots dy_1 d\omega_k\ldots d\omega_0 d\nu . \tag{6.4}$$

The term in the integrand $B^{(k)}$ represents the path that a ray has followed from starting at point y with angle ω_0 to arriving at point x. It is explicitly given in Equation (5.59) of the last chapter. The term $P(y,\omega_0,\nu)$ gives the probability that a ray emitted from point y will leave with angle ω_0 and frequency ν.

The integrals in the solution can of course be evaluated with any appropriate integration method. But the high dimension leaves in practise only the Monte Carlo method (see appendix for a general description of Monte Carlo integration). The interpretation of the calculation of Equation (6.4) with the help of Monte Carlo ray tracing is clear. First, the initial direction and energy of the ray are sampled from the distribution $P(y,\omega_0,\nu)$. The integrand $B^{(k)}$ is then evaluated according to the scattering events which occur along the ray path. Averaging the results of the kth scattering event at point x over many rays which originated at point y gives an estimate of $\widetilde{K}_T^{(k)}(x,y)$.

As mentioned in Section 5.4.1, the surface D is usually discretized into surface elements D_i. The quantity of interest is then the total energy transferred from the originating surface element D_O to the target element D_T. This may be described by defining

$$\Phi_T(D_O,D_T) = \frac{1}{Area(D_T)} \int_{D_T} \int_{D_O} \Phi_T(x,y)dy\,dx . \tag{6.5}$$

The surface elements should be small enough that the temperature can be considered as approximately constant, but large enough that the number of rays emerging is significant. The distribution of originating points in the surface element is uniform under the assumption of a homogeneous temperature. The positions can either be obtained by sampling from a uniform distribution or from a regular grid.

The numbers of the rays emitted from a surface element should be carefully chosen to minimize the variance of the calculation (see Equation (A.16) for a simple example of this principle). Let $E_i = \int_{D_i} E(x)\,dx$ be the flux emitted from the surface element i, where $E(x)$ is defined by Equation (5.60). If N_{total} is the total number of rays available for the Monte Carlo calculation, then the optimal number of rays for surface i is proportional to the flux

$$N(i) = \frac{E_i}{\sum_{j=1}^{N} E_j} N_{total} . \qquad (6.6)$$

We consider now how to sample from the function $P(y, \omega_0, \nu)$. The steric angle ω_0 may be written in terms of coordinates ϑ and φ with $0 \leq \vartheta \leq \pi/2$ and $0 \leq \varphi \leq 2\pi$ so that the differential solid angle $d\omega_0$ is $d(\cos \vartheta)d\varphi$. In these coordinates, the normal vector to the surface is $(0, 0, 1)$. According to Equation (5.59) the function P may be expressed as

$$P(y, \cos \vartheta, \varphi, \nu) = \frac{e_{\nu,T}(y, \omega_0) c \, i_{\nu,T}^{BB}(y)}{e_T^{eff}(y) \sigma T^4(y)} \cos \vartheta \qquad (6.7)$$

and

$$\int_0^\infty \int_0^{2\pi} \int_0^{\frac{\pi}{2}} P(y, \cos \vartheta, \varphi, \nu) d(\cos \vartheta) d\varphi d\nu = 1 .$$

When rays are distributed according to the the probability density $P(\cos \vartheta, \varphi, \nu)$, they define the nodes for the Monte Carlo integration over the variables $\cos \vartheta$, φ and ν. The question is then how to sample values of these variables such that they are distributed according to the density function P. This of course depends on the nature of the surface. In general, the angular dependence of the emissivity changes with the wavenumber ν. Moreover, if the surface is anisotropic, this dependence may be complicated. Depending on the functional form of P, various Monte Carlo sampling methods may be applied.

To illustrate, we proceed with the simplest case of an isotropic emissivity which has no angular dependence. This approximation arises when $e_{\nu,T}(y_0, \omega_0)$ is replaced by the hemispherically averaged emissivity $e_{\nu,T}^{hem}(y_0)$. The densities of the variables $\cos \vartheta$, φ and ν are then independent, and the density function may be factored as

$$P(\cos \vartheta, \varphi, \nu) = P_1(\cos \vartheta) P_2(\varphi) P_3(\nu) , \qquad (6.8)$$

where

$$\begin{aligned}
P_1(\cos \vartheta) &= \cos \vartheta , & \vartheta \in \left[0, \frac{\pi}{2}\right] \qquad (6.9) \\
P_2(\varphi) &= 1 , & \varphi \in [0, 2\pi] \\
P_3(\nu) &= \frac{2h\pi}{c^2 e_T^{eff} \sigma T^4} \frac{\nu^3}{e^{h\nu/k_B T} - 1} , & \nu \in [0, \infty] .
\end{aligned}$$

Sampling the angular variables is now easily accomplished using the direct method described in the appendix. In this case, if $R_{\cos \vartheta}$ and R_φ are uniformly distributed, independent random numbers on the interval $[0, 1]$, then the sampled angular variables are given by

$$\cos \vartheta = \sqrt{R_{\cos \vartheta}} \tag{6.10}$$
$$\varphi = 2\pi R_\varphi .$$

Sampling the spectral variable requires slightly more work because the function $P_3(\nu)$ does not have a closed form anti-derivative which may be easily inverted. Typically this difficulty is overcome by replacing $P_3(\nu)$ by a piecewise constant approximation on the n spectral intervals

$$\{[\nu_0, \nu_1], [\nu_1, \nu_2], \cdots, [\nu_{n-1}, \nu_n]\}$$

ν_0 and ν_n are chosen so that $P_3(\nu)$ is essentially zero for smaller or larger values. In this approximation $P_3(\nu) = P_3^i$ for $\nu \in (\nu_{i-1}, \nu_i)$ where

$$P_3^i = \frac{e_{i,T}^{hem} I_{i,T}^{BB}}{\nu_i - \nu_{i-1}} \tag{6.11}$$

and

$$e_{i,T}^{hem} = \frac{\int_{\nu_{i-1}}^{\nu_i} e_{\nu',T}^{hem} \frac{2h}{c^2} \frac{\nu'^3}{e^{h\nu'/k_B T}-1} d\nu'}{\int_{\nu_{i-1}}^{\nu_i} \frac{2h}{c^2} \frac{\nu'^3}{e^{h\nu'/k_B T}-1} d\nu'} , \quad I_{i,T}^{BB} = \int_{\nu_{i-1}}^{\nu_i} \frac{2h}{c^2} \frac{\nu'^3}{e^{h\nu'/k_B T}-1} \tag{6.12}$$

The associated cumulative distribution function is then piece-wise linear, and may be easily inverted to give sampled values of ν.

The distribution function of Equation (6.7) is for radiation emitted from a (locally) planar surface. There is another, qualitatively different kind of emitting surface which should be mentioned. This is the tungsten filament. Again we consider the simplest case of an isotropic, angle-independent emissivity. Let $\vec{t}$ be the tangent vector of the filament and $\vec{n}_t$ be orthogonal to $\vec{t}$. Let ω be the direction of a ray starting from the filament, defined by ϑ, the angle between ω and $\vec{t}$, and φ, the angle between the $\omega - \vec{t}$ plane and $\vec{n}_t$. In this case the angular variables are again independent and have probability distribution functions

$$P_1^{fila}(\cos \vartheta) = 1 \quad , \quad \vartheta \in \left[-\frac{\pi}{2}, \frac{\pi}{2}\right] \tag{6.13}$$
$$P_2^{fila}(\varphi) = 1 \quad , \quad \varphi \in [0, 2\pi] .$$

In this case the sampled angles are given by

$$\cos \vartheta = 2R_{\cos \vartheta} - 1 \tag{6.14}$$
$$\varphi = 2\pi R_\varphi .$$

The Equations (6.10) may be regarded as a change of variables which maps the hemisphere defined by (ϑ, φ) to the unit square defined by

(R_ϑ, R_φ). This change of variables (or its equivalent in the case of non-isotropic, angle dependent reflectivities and emissivities) may be applied to all the angular integrations of Equation (6.4). In terms of the simulation, this means that the same procedure just outlined may be used to compute all non-specular reflections of a ray, as well as its initial direction. In the case of specular reflection (see Section 5.4.2) there is only one angular integration for the initial direction. For diffuse reflection (see Section 5.4.3) Equations (6.10) are used to sample the direction of each reflection. For a realistic surface with some roughness, the angular dependence of the reflection will be neither diffuse nor specular. There are modeling efforts for the description of such reflection functions in the field of computer graphics. See [98] for a such a model suited for Monte Carlo simulation.

The question of how to handle the absorptivity, reflectivity and transmissivity remains to be answered. This was already addressed somewhat in Section 5.4.2, where a probabilistic alternative to ray splitting was offered when semi-transparent material are present. There are two options for a Monte Carlo simulation which are based on the choosing either a deterministic or probabilistic view of the process.

The first option is the deterministic approach known as the fractional absorption method. The physical picture is that of a ray which looses a fixed percentage of its energy $a_{\nu,T}(x, \omega)$ at each surface interaction due to absorption. The ray is transformed into a reflected ray with the fraction $r_{\nu,T}(x, \omega)$ of the original energy, and in the case of a semi-transparent material, a transmitted ray with the fraction $t_{\nu,T}(x, \omega)$ of the original energy. Conservation of energy is expressed by the fact that $a_{\nu,T} + r_{\nu,T} + t_{\nu,T} = 1$. The rays continue on through an infinite number of surface interactions. In practise, a ray will lose all but a negligible percentage of its original energy within a finite number of reflections, after which its path may be terminated. It should be noted that in all but the specularly reflecting case, the directions of the reflected rays will still be chosen randomly in the Monte Carlo simulation.

The second option is the probabilistic approach known as the discrete absorption method. This is described by Equation (5.72) for specular reflection and Equation (5.75) for diffuse reflection. The physical picture here is that of a ray which is either completely absorbed, reflected or transmitted. If absorbed, the path is terminated, and its total energy is added to the surface. If reflected or transmitted, the ray continues on as a single ray with no loss of energy. The decision of whether the ray is absorbed, reflected or transmitted is a random choice based on the probabilities $a_{\nu,T}$, $r_{\nu,T}$ and $t_{\nu,T}$.

As with all integration problems, the dimension of the domain is of crucial importance. Table 6.1 summarizes the dimension of the integration domain for the two methods described here. For simplicity, it is assumed that all surfaces are opaque ($t_{\nu,T} = 0$). The first integration is over the

method	surface	pos.	ener.	dir.	k^{th}refl.	k^{th}dir.	total
fractional	specular	1	1	2	0	0	4
	not specular	1	1	2	0	2	4+2k
discrete	specular	1	1	2	1	0	4+k
	not specular	1	1	2	1	2	4+3k

Table 6.1: Dimension of Integrals in Fractional and Discrete Absorption Methods from the sampling of position, energy and direction at the origin of the ray and the reflection and direction of the kth encounter with a wall.

original position of the ray. The dimension depends on the spatial symmetry and dimension of the surface; it is one for a 3 dimensional, axisymmetric problem. The next integration concerns selecting the direction and energy of the ray, and thus requires 3 dimensions. Except in the case of specular reflection, each subsequent surface reflection adds 2 dimension which are used to determine the new direction of the ray. No additional dimensions are required to determine energy deposition if the fractional absorption method is used; however, the discrete reflection model adds one more dimension per reflection for this decision.

It is important to note that although the dimension for the discrete absorption method is higher than for the fractional absorption method, the average computational time required to compute a ray path may be substantially shorter due to the fact that once a ray is absorbed, its path is terminated. This allows a larger sample size for the discrete method, which may offset the penalty associated with the higher dimensionality.

6.2   Quasi-Monte Carlo Methods

The radiation transport treated here is a linear phenomenon. This means that the radiation scatters off a fixed background and the presence of one ray does not influence the paths of other rays. This is in sharp contrast to transport in a rarefied gas where it is the particle interactions which drive the flow. Also for electron transport non-linear effects are important, even if electron-electron collisions are insignificant. The electron path is influenced by the self induced electric field.

The linearity of many radiation problems leads to greater flexibility in the analysis and numerical methods used to solve them. In ray tracing whether each ray path is traced to completion in a serial fashion before starting the next or many rays are followed simultaneously in a parallel fashion does not affect the results of simulations using random sequences for event decisions. Many Monte Carlo techniques like stratification have been developed to handle various kind of problems. These mainly address the question of variance reduction [31]. This section focuses on a lesser known technique, quasi-Monte Carlo, which has been shown quite effective for integration and certain radiation transport problems.

The key idea behind quasi-Monte Carlo is that for linear simulations, because the particle or ray paths do not influence each other, the numbers used to determine the paths do not need to be independent. It is only necessary that the paths be selected uniformly from the space of all possible paths so as to obtain a representative sample of particle or ray behavior. Each path may be thought of as a point in a multi-dimensional unit cube, with the individual dimensions corresponding to the scattering decision making parameters. If the points corresponding to the simulated paths are well distributed throughout the cube, then the simulation should accurately model the physical behavior of the system. Quasi-Monte Carlo provides a means of selecting a more uniformly distributed set of paths than random numbers. However, the independence of the paths is lost. For linear steady simulations which are of interest here, this is not important. For non-linear or time dependent problems, great care must be taken to avoid difficulties which arise from correlations [65], [55].

These ideas are more easily understood by considering the simpler problem of integrating a function $f(x)$ over the d-dimensional unit cube I^d. As discussed in the appendix, the Monte Carlo estimate of the integral is given by

$$\int_{I^d} f(x) \, dx \approx \frac{1}{N} \sum_{i=1}^{N} f(x_i) \tag{6.15}$$

where $\{x_i\}$ $i = 1, \ldots, N$ is a sequence of independent, uniformly distributed random points in I^d. The error in this approximation, ϵ, satisfies the equality

$$E(\epsilon^2) = \frac{\sigma^2(f)}{N} \tag{6.16}$$

where $E(\cdot)$ is the expectation with respect to the random points and $\sigma^2(f)$ is the variance of the integrand. This shows that the error in the approximation converges like $N^{-1/2}$.

The independence of the sequence $\{x_i\}$ plays only a secondary role in the evaluation of the integral. It ensures that, should additional points (i.e., a larger N) be required, the entire set will still be uniformly distributed throughout the cube. It is, however, the uniformity of the points that ensures convergence. This suggests that if there were a sequence of points which had better uniformity properties than a randomly chosen sequence, the resulting error would be smaller.

This leads to a number of questions such as what is a qualitative way of measuring uniformity, do sequences with greater uniformity than random exist, and how is the uniformity of a sequence related to the integration error when the sequence is used to estimate the integral. The answers to these questions form the basis of the theory of quasi-Monte Carlo methods.

6.2.1 Discrepancy

In order to measure the uniformity of a sequence of N points, the idea of discrepancy was introduced. For simplicity, only sequences in the unit cube will be considered with their uniformity measured against the uniform measure dx.

For any subrectangle J of the unit cube (i.e., a parallelepiped with sides parallel to the coordinate axes) and any sequence of N points $\{x_i\}$ in the cube, define

$$R_N(J) = \frac{1}{N} \sum_{i=1}^{N} \chi_J(x_i) - m(J) \qquad (6.17)$$

Here χ_J is the characteristic function of J and $m(J)$ is its volume (i.e., $m(J) = \int_{I^d} \chi_J(x)\, dx$). The L_∞ discrepancy of the sequence $\{x_i\}$ is then defined as

$$D_N = \sup_{J \in E} |R_N(J)| \qquad (6.18)$$

where E is the set of all subrectangles of I^d.

$R_N(J)$ is a comparison of the percentage of points of the sequence which lie in J to the percentage of the volume of the unit cube occupied by J. A uniform sequence would be one which puts the right percentage of points, relative to the volume, in every set J. The L_∞ discrepancy gives the worst case relative to rectangles. Thus if D_N is small, the sequence should be uniform. In fact, an infinite sequence is defined as uniform if $\lim_{N \to \infty} D_N = 0$, where D_N is the discrepancy of the first N terms of the sequence. It follows from the Central Limit Theorem that a random sequence with probability density 1 is almost surely a uniform sequence. With the law of iterated logarithms, it can be shown that the discrepancy of a random sequence satisfies

$$D_N = \mathcal{O}\left(\sqrt{\frac{\log \log N}{N}} \right) \qquad (6.19)$$

almost surely. Thus the discrepancy of random sequence shows similar lack of dimensional dependence and convergence rate as the error in a random Monte Carlo integration.

6.2.2 Quasi-Random Sequences

From the definition it follows easily that a lower bound for the discrepancy of any sequence is $1/N$. The next question is whether there exist sequences

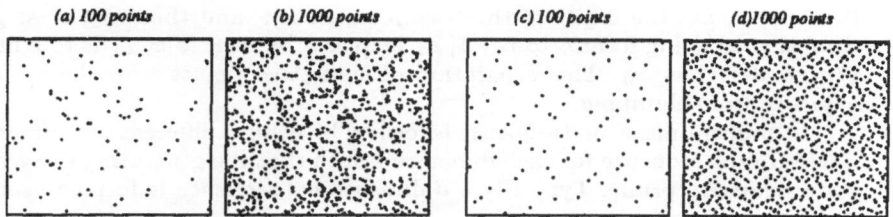

Figure 6.1: Coverage of a square with points generated from a pseudo-random (pictures (a) and (b)) and a quasi-random sequence of Halton type (pictures (c) and (d)).

which come closer to achieving this lower bound than a random sequence. The answer is yes, and sequences which have optimal discrepancy bounds are known as quasi-random, or low discrepancy sequences.

Before considering such sequences, it is worthwhile to examine the case of regular lattice points (grids). It may initially seem that this would provide optimal uniformity; however, particularly in high dimensions, a number of problems arise which reduce their effectiveness. First, the discrepancy of a rectangular lattice of N points is $N^{-1/d}$. Second, to refine the lattice requires increasing the number of points by a factor of 2^d. Finally, for high dimensions, it is usually impossible to put down enough lattice points to get good resolution.

Quasi-random sequences combine the advantage of random sequences (points can be added incrementally while still maintaining uniformity) with the advantage of a lattice (points are well separated and do not clump together). This is achieved in a variety of ways, depending on the sequence, including using different grid spacing for each dimension (the Halton sequence), or only partially filling out a lattice structure before starting to fill in the next level of refinement (the Sobol' and Faure sequences).

For problems of low to moderate dimension (say ≤ 10), the Halton sequence has been found to be slightly more effective. This includes many radiation transport problems. Thus as an example, the Halton sequence will now be described in more detail. A variety of other sequences can be found in the literature [73],[88], [27], [74].

A Halton sequence in one dimension is generated by choosing an integer p and expanding the sequence of integers $\{1, 2, \ldots, N\}$ in base p. The nth term of the Halton sequence x_n is given by

$$x_n = \frac{a_1}{p} + \frac{a_2}{p^2} + \cdots + \frac{a_k}{p^k} \tag{6.20}$$

where

$$[n]_p = a_k a_{[k-1]} \cdots a_2 a_1 . \tag{6.21}$$

Here the a_i are the digits of the base p expansion, and thus $0 \leq a_i < p$. Note that in going from n to $n + 1$, a_1 increases by one, so x_{n+1} is roughly $1/p$ greater than x_n. This separation of successive points helps keep the sequence from clumping.

A multi-dimensional sequence is formed by using a different one dimensional Halton sequence for each dimension, such that the generating integers p are relatively prime. Typically a d-dimensional sequence is formed using the first d primes.

The discrepancy of the Halton sequence has been shown to satisfy

$$D_N \leq C_d \frac{(\log N)^d}{N} + \mathcal{O}\left(\frac{(\log N)^{d-1}}{N}\right) \qquad (6.22)$$

where

$$C_d = \prod_{k=1}^{d} \frac{p_k - 1}{\log p_k} \qquad (6.23)$$

and the p_k are the generating primes. Up to this constant, this is regarded as the best bound possible for any sequence. Clearly for large enough N, the discrepancy of the Halton sequence is smaller than the expected discrepancy of a random sequence. Numerous other sequences have been suggested for which similar bounds on discrepancy have been shown.

6.2.3   Integration and Discrepancy

The standard justification for using quasi-random sequences as numerical integration nodes is the Koksma-Hwalka inequality, which relates integration error to the discrepancy of the sequence used. This is described below. However, as discussed in [66] and [67], this inequality and the bounds on discrepancy lead to vast over estimates of the integration error for the range of N which is of interest. Thus the bounds and the inequality are mainly of theoretical interest. The true behavior of quasi-Monte Carlo methods has only been established through computational experiments such as described in [67], [17], [46], [76], [85].

For completeness the theoretical results will now be briefly described. A full description can be found in [75]. The Koksma-Hwalka inequality states that for functions with bounded variation $V(f)$ in the sense of Hardy and Krause, the error ϵ of the Monte Carlo estimate of the integral of f obtained using a given sequence of N terms with discrepancy D_N satisfies

$$\epsilon < V(f) \, D_N \qquad (6.24)$$

For smooth functions which vanish on the boundary of the unit cube, the variation is given by

$$V(f) = \int_{I^d} \left| \frac{\partial^d f}{\partial x_1 \cdots \partial x_d} \right| dx_1 \cdots dx_d \qquad (6.25)$$

The results of [67], however, show that variation is a poor indicator of the ease with which a function f may be integrated.

Another theoretical result discussed in [100] and [66] considers a space of functions equipped with a Wiener measure. These functions are continuous and said to be half differentiable. The result is that for a fixed sequence, the expected integration error over all the functions in the space with respect to the Wiener measure is the L_2 discrepancy of the sequence. (The L_2 discrepancy is similar to D_N except that the sup $|R_N(J)|$ taken over all rectangles J is replaced by integrating $R_N^2(J)$ over all rectangles). This result is extended to differentiable functions with a Wiener sheet measure in [79].

In practise the size of the integration error associated with a quasi-Monte Carlo approximation generally follows several heuristics, but no hard estimates are available. This is different than the random case, where the expected error is given explicitly.

It is worthwhile to first consider the similarities between standard random Monte Carlo and quasi-Monte Carlo before examining the potential improvements that quasi-random sequences may offer. For random Monte Carlo, the error is statistical in nature and convergence is given only in expectation. The Central Limit Theorem ensures that for large enough N, the probability of a large error is small. However, for any given set of N randomly chosen points, the error may be large. The variance of the integrand provides a measure of the size of the statistical scatter which may be expected.

Quasi-random sequences also lead to results which appear to be statistical in nature, although there is no underlying probability distribution function, so this is not a well defined concept. The theory provides only an upper bound on the error for a fixed integral. In practise, using a different contiguous subsequence of length N from the same (infinite) quasi-random sequence will produce a scatter of estimates for the integral, despite the fact that the discrepancies of these subsequences are almost identical. This scatter is not surprising, as each subsequence evaluates the integrand at different points. It is therefore not practical to speak of the error associated with using N terms of a quasi-random sequence except either as related to an upper bound, or more importantly, as a kind of expectation.

Despite the lack of a direct theoretical underpinning for the expectation of a quasi-random sequence, the idea has proven nonetheless the most useful for describing the behavior of quasi-Monte Carlo methods. A typical computational used to determine the convergence rate in N for a given integrand is conducted as follows. The integral is estimated using N contiguous points of a quasi-random sequence. This is repeated M times (with M usually between 10 and 100) for this fixed N, and the root mean square error associated with these trials is computed. The procedure is then repeated using a larger value of N, using a completely separate subsequence of the quasi-random sequence to avoid correlations as much as possible among the various values of N. It is found that the expected error may be fit to a function of the form cN^β. Thus the values of N are usually chosen to be logarithmically spaced to aid in parameter evaluation. For random sequences, the expected value of β is -0.5, and the constant is the standard deviation of the integrand (the square root of the variance).

The basic result is that quasi-Monte Carlo is almost always better than random Monte Carlo for evaluating integrals. The improvement is both in the rate of convergence β and in the size of the error (the constant c). The degree of improvement depends on the integrand. It is only possible to report general trends, to which there are always exceptions.

As suggested by the bound on discrepancy, there is often a decay in the convergence rate as the dimension of the problem increases. This may be attributed to the difficulty in filling out a high dimensional space uniformly. The tendency to have errors similar to random is often seen in high dimensions. Computations of discrepancy suggest that only after a transition value of N, exponentially large in dimension, does the quasi-random improvement appear. This is tempered by the fact that many nominally high dimensional problems have only limited dependence on the upper dimensions. Thus greater uniformity in the first few dimensions may lead to significant error reduction.

It should be noted that low dimensional integrals are often better handled by quadrature formulas. The success of this approach depends, however, on the smoothness of the integrand. Smoothness does not directly affect Monte Carlo or quasi-Monte Carlo methods, so that using a quasi-random sequence to evaluate even a two dimensional integral may give the lowest error.

While differentiability does not affect the quasi-Monte Carlo performance, it has been observed that lack of continuity may degrade the results. Again, there is an improvement over random Monte Carlo, but it may be significantly reduced for step functions as opposed to continuous functions. This is of particular importance to simulation problems because decision processes are often described by integrals of step functions. The reason for the decrease in performance may be that points near the boundary of the step have essentially a random chance of landing inside or out of the step region. This additional randomness drive the error towards $N^{-.5}$ convergence. This effect is particularly noticeable in high dimensions because the ratio of surface area to volume increases with dimension.

Typically convergence rates for quasi-random integration lie between $N^{-.5}$ and N^{-1} and the actual error is a factor 2 to 10 smaller than the random error for the range of N usually considered. Because of the square root convergence of random methods, a factor of 3 improvement corresponds to an order of magnitude improvement in computation time. This is clearly a significant savings, given that quasi-Monte Carlo methods often have virtually no additional computational costs over the standard method.

Other than simple integration, a number of other Monte Carlo applications have been attacked with quasi-Monte Carlo methods. In [65] a quasi-random walk is used to solve the heat equation, while in [55] the spatially homogeneous Boltzmann equation for gas dynamics is considered. Radiation absorption in an atmosphere is examined in [76]. In the next chapter we present a numerical study of an idealized RTP reactor which demonstrates both the application of Monte Carlo methods to a ray tracing problem as well as the advantage of using quasi-Monte Carlo methods.

6.3 Coupling Radiation and Gas Transport

In most micro-electronic processing applications, radiative heat transfer is only one of many physical processes. Effective equipment simulation requires a smooth coupling of the radiation model to a state of the art CFD/CVD simulator for fluid flow and chemistry. Popular examples of CFD/CVD simulators are the finite volume codes FLUENT [2] and PHOENICS/ACCESS-CVD [3] and the finite element code FIDAP [1]. An extensive description of CVD models in CFD finite volume codes is given in [50]. A Monte Carlo radiation simulation similar to what is described above has been coupled to the PHOENICS/ACCESS-CVD code for the case of axisymmetric geometries. We review now a few of the key elements of this coupling that allow the simulation to run efficiently.

The first question is how to efficiently trace a ray path through a complicated reactor. This is essentially the same problems as the free molecular flow of a gas molecule, which was discussed in Chapter 3. The simplest approach is to use the underlying computational grid for the CFD simulator, which must in any case be provided. The ray tracing may then be done from face to face of the grid cells. Just as with a rarefied gas, however, this may be inefficient if there are large interior regions with no boundaries. In this case, the idea of the macro-cells (see Section 3.1.5) is useful. Here a coarse grid is laid over the CFD grid; rays are then tracked across the macroscopic cells.

There may be, however, exceptional regions where the ray tracing cannot be done solely through the cells of the CFD grid because the grid cannot sufficiently resolve some geometric detail, important only for radiative heat transfer, in a reasonable way. An example of such a region is a lamphouse with an array of several lamps and parabolic reflectors. Even a finely discretized CFD grid could only approximately describe the filament and the reflector shape. For such regions it is usually more efficient and accurate to include a special subroutine, written specifically for the given geometry, to handle the ray tracing. The subroutine may exploit facts such as that the surface is made up of intersecting parabolas (in the case of the lamphouse). Often there will be a set of simple parameters, such as curvature and orientation, which may be more easily adjusted than changing a whole grid.

The second issue of importance is how to link the results of the heat transfer calculation to the fluid computation. The net radiation heat flux $R(x)$ (see Equation (5.61)) is used as a boundary condition for the fluid energy balance equation. Typically, an initial estimate of the radiation temperature is given. From this the radiation exchange matrix $\Phi_{T_{ref}}$ is computed. The radiation simulation then determines $R(x)$ and the temperature distribution on the surfaces. This information is used to compute the fluid flow. The calculation may result in a new surface temperature distribution, as well as modified surface properties due to thin film surface coating. It is then necessary to recompute $\Phi_{T_{ref}}$ with the updated information. For a transient calculation, it is not possible to calculate Φ_T after each time step. In this case it is particularly important to capture at least the radiation temperature dependence of Φ_T given by approximation (5.66) so as to compute the radiation exchange matrix as few times as possible.

Chapter 7

Radiation Transport Simulations

The theory and numerical methods discussed in the preceding chapters for radiative heat transfer will now be illustrated with the help of three examples. In the first case, we consider an idealized RTP reactor in order to focus on some numerical questions and determine the effectiveness of quasi-Monte Carlo methods. The second case involves the physically realistic simulation of an existing industrial reactor. The necessity of accurately modeling all relevant aspects of the equipment is illustrated here by comparing several simulations, based on different models, with experimental results. Finally, results are given for a third reactor for which the details of the lamphouse heating chamber have been computed by a special subroutine separate from the flow simulator geometry, as discussed in Section 6.3.

7.1 Case Study of an RTP Reactor

[1] The calculation of the heat transfer in a simplified RTP reactor will now be discussed in order to study some numerical questions which arise in the simulation. As mentioned in the last chapter, there are always two approaches for handling the surface scattering events which arise in computation of radiative heat transfer: the fractional absorption method and the discrete absorption method. We consider here which of these methods is more computationally effective. We also demonstrate the advantage of using quasi-Monte Carlo methods for this problem and consider the effect of quasi-random sequences on the question of fractional or discrete absorption.

---

[1] Portions of this section are reprinted from [46] by courtesy of Marcel Dekker, Inc.

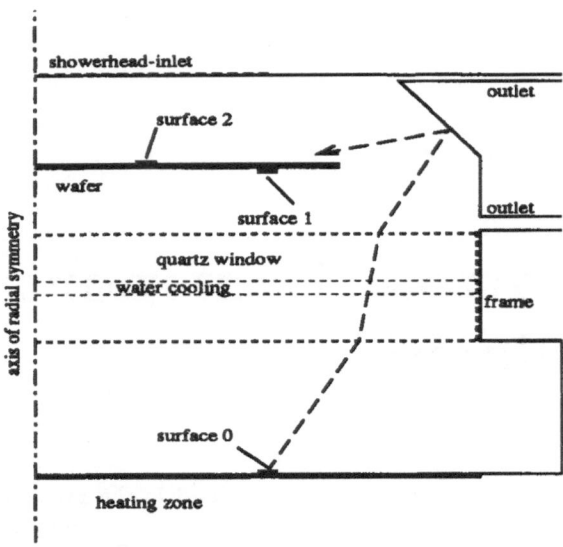

Figure 7.1: Geometry of the reactor and position of specific surface elements.

7.1.1 A Simplified Radiative Heat Transfer Reactor

A typical application of rapid thermal processing is to use thermal radiation to heat a silicon wafer placed inside a reactor. The control of the temperature distribution of the wafer in such processing equipment is of prime importance. Since the in situ control of the temperature is a partially unsolved problem, the simulation of the physical mechanisms determining the temperature cycles can provide important information for the optimization of such processes. One of the problems which can be solved by such a simulation is high accuracy modeling of the radiative heat transfer from the heater to the wafer. Figure 7.1 shows the draft of the cylinder symmetric projection of a typical single wafer reactor. Because of the cylinder symmetry, the surface elements shown in the figure actually represent annuli. The heater, labeled *surface 0*, had an inner radius of 5.5 *cm* and an outer radius of 6.0 *cm*. When a point source heating element was used, it was placed at a radius of 5.75 *cm*. The annulus of the wafer labeled *surface 1* was placed directly above the heater and had an area of 18.06 *cm²*. *Surface 2*, on the back side of the wafer, had an inner radius of 2.5 *cm* and an outer radius of 3.0 *cm*, corresponding to an area of 8.64 *cm²*.

The mathematical formulation of this problem has been described in the previous two chapters. For this example we will make the following assumptions. First, all surfaces will be considered opaque, so there is no transmissivity. Second, the reflectivity and absorptivity will be constant, independent of temperature and the spectral frequency (the wafer and the walls, however, may have different values for these constants). Finally, the only source of radiation entering the problem will be from the heater.

7.1.2 Numerical Experiment

The typical numerical experiment consisted of emitting N rays from a surface or point source located at *surface 0*. The point source is actually a circle under cylinder symmetry. For the surface source, the rays were distributed sequentially by increasing radius on a one-dimensional grid so the distribution on the corresponding annulus would be uniform. The initial direction was sampled using Lambert's Law, which means that the zenith angle was sampled from a cosine distribution, while the azimuthal angle was sampled from a uniform distribution on $[0, 2\pi]$. The sampling was done using one point from a multi-dimensional quasi- or pseudo-random sequence, such that two angles were assigned separate dimensions. A third dimension was used to sample the initial energy of the ray from a modified version of Planck's black body distribution. The fractional method then consisted of tracing the ray path through the reactor according to the laws of specular reflection, such that at each reflection a certain percentage of the ray's energy, given by the absorptivity at that point, was transferred to the surface element in question. The tracing continued until the ray had lost a fixed percentage of its initial energy. The discrete method varied from the fractional method only in that at each reflection, either no energy was absorbed and the ray continued on, or the entire initial energy was absorbed and the ray tracing was terminated. In both cases, the final answer was computed as the percentage of the energy which left the source that was absorbed by the target element.

The ultimate goal of the experiments was to determine the accuracy as a function of computation time. This is discussed in more detail below. The first step towards this goal was to compute error size as a function of N, the number of rays emitted from the source. The approach taken here was to compute the "expected" convergence rate and error size by performing the calculation 30 times for each value of N using different, "independent" subsequences of the sequences in question. The error was then averaged to give a better estimate of what the expected error using N rays would be. This calculation was then repeated using a larger value of N, again using different subsequences. In this way, errors at different values of N are also independent, which allows for an unbiased description of the convergence rate. The values of N tested were chosen to be evenly spaced on a logarithmic scale, so that convergence of the type N^β could more easily be studied.

The terms expectation and independence are only strictly defined for a sequence of random variables. Performing the experiments described above with such a random sequence would show converge at a rate of $N^{-.5}$ and an error size determined by the variance of the integrand. The results for the pseudo-random sequence do indeed show a convergence rate of approximately $1/\sqrt{N}$. The same method of computing expected error is then extended to the quasi-random sequences as a means of comparison. It should be noted that this kind of convergence study using averaging and independent N's requires considerably more computation time than a straight forward simulation using a fixed number of rays.

An additional difficulty in this convergence study was that the exact solution for the energy transfer was not known. Thus it was necessary to estimate the exact answer from the Monte Carlo calculation and use this to measure the error at each value of N. The estimate for the exact answer was obtained by first averaging the 30 results obtained for each of the last four values of N, then taking a weighted average of these four values. Under the assumption of a Gaussian error distribution, it can be shown that this gives an optimal estimate to the true answer.

As stated above, the real quantity of interest is computation time to reach a specified error tolerance. This is of particular importance in comparing the fractional and discrete methods. The discrete method requires fewer reflections, but at the cost of being a higher dimensional integral. Thus N might be larger in the discrete case, but the time to compute each ray path would be smaller than in the fractional case. The experiments just described address only the question of convergence with N. The problem is that computation time depends on the coding of the programs and sequences and is very machine dependent. To avoid issues of optimal coding, which would in any case only be accurate for a specified machine, we devised the following means of comparison for the fractional and discrete methods.

The computation consists roughly in computing an average of n_r random numbers in a time $n_r t_r$ and performing an average of m ray tracing steps in a time $m t_t$. Thus the total computing time is $T = N(n_r t_r + m t_t)$. We observed that most of the time is spent on ray tracing. Because our geometry is not very complex, the most realistic case will be a situation very close to the limit in which all computation time is spent on tracing. Then a comparison T^f/T^d of the computation time with fractional and discrete method will only depend on the ratio m^f/m^d of the number m^f of ray traces with the fractional method to the average number m^d of ray traces with the discrete method. It is simple to realize that $m^d = 1/A$, where A is the absorptivity. However, m^f depends on the desired accuracy of the calculation. If we require that a ray be traced using the fractional method until the ratio of its final energy to its initial energy is less than α (in our case $\alpha = 10^{-4}$), we have $m^f = [\log(\alpha)/\log(1 - A)] + 1$. With our energy limit and $A = 0.4$ we get $T^f/T^d = 7.6^2$. This gives the scale to allow the

---

[2] In our actual calculation this limit was between 6.23 and 6.81 for the different number

Surface	Abs.	Converg. $\beta : N^\beta$	Error at $N = 10^5$	Improv. Factor	Error at Time T^*	Improv. Factor
1	0.1	-.64	.0027	2.3	.0027	2.3
	0.4	-.70	.0033	3.4	.0012	4.6
	0.7	-.70	.0042	3.7	.0008	6.0
	mix	-.66	.0036	2.8	.0017	3.3
2	0.1	-.64	.0049	2.7	.0049	2.7
	0.4	-.65	.0099	2.4	.0036	3.0
	0.7	-.65	.0152	2.4	.0032	3.4
	mix	-.65	.0097	2.3	.0047	2.7

Table 7.1: Convergence results for fractional absorptivity method using the Halton sequence.

comparison of the computation time of the fractional and discrete method.

Three absorptivities were tested, $A = 0.1, 0.4$ and 0.7. The associated number of reflections m^f necessary using the fractional method to achieve the accuracy $\alpha = 10^{-4}$ were 88, 19 and 8 respectively. In order to compare the fractional method across absorptivity, a reference time T^* was chosen as the time required to compute 10^5 rays with $A = 0.1$ using the fractional method. Under the assumption that ray tracing dominates the computation time, $88/19 = 4.63$ times as many rays could be calculated for $A = 0.4$ in the same T^*, and $88/8 = 11$ times as many rays for $A = 0.7$. A similar scaling is used to determine the number of rays which can be computed in T^* for the different absorptivities using the discrete method.

7.1.3 Numerical Results

For the sake of brevity, results are presented here for only one quasi-random sequence, Halton. The Sobol' sequence was also tested, but the results were generally the same as for Halton; on the occasions where there was a difference, Halton appeared slightly better. Also, all calculations described here were performed using a point source for the radiation. Experiments were also done using a surface source; however, as there is no significant difference in the results of the two approaches, only the point source results are given here.

The first set of calculations is illustrated in Figure 7.2 and Table 7.1. This figure shows the expected relative error $\epsilon(N)$ as a function of the

generators. Hence we were pretty close to the assumed limit.

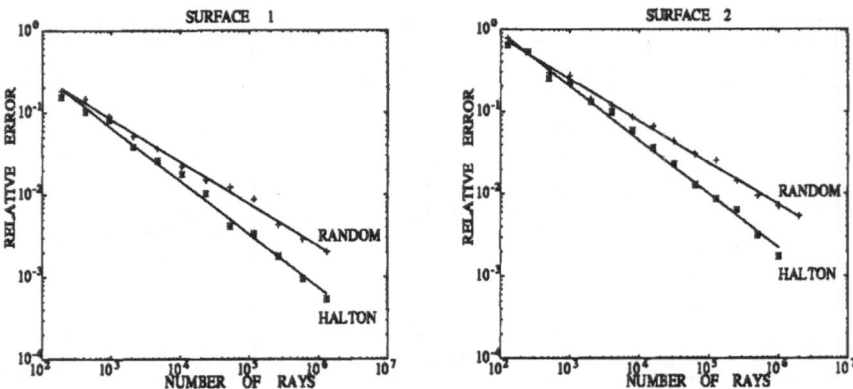

Figure 7.2: Comparison of random and quasi-random sequences using the fractional absorption method with absorptivity = 0.4.

number of emitted rays N for the Halton sequence and a pseudo-random sequence using the fractional absorption method with a constant surface absorptivity of 0.4. Results are given for *surface 1* on the wafer (see Figure 7.1), which is directly visible to the source, and for *surface 2*, on the back side of the wafer. The plotted points are the calculated errors for various N (averaged over 30 runs), while the lines are a least squares fit of the data to the functional form

$$\epsilon(N) = c\,N^{\beta}\;. \tag{7.1}$$

This is the correct form for the expected error using a random sequence, in which case c is the standard deviation of the integrand and β is $-1/2$. As plotted on a log scale, the error appears as a line with slope β.

Figure 7.2 illustrates a clear advantage of using a quasi-random sequence over a pseudo-random sequence in the calculation, both in error size and in convergence rate (i.e., β). The error in calculating the energy transfer to *surface 1* with $N = 100000$ is over a factor of three smaller if the Halton sequence is used than if a random sequence is used. This figure also shows that the accuracy of the calculation depends on the region of the wafer in question. The error in computing the energy transfer to *surface 2* is considerably higher than for *surface 1*. This is related to the fact that a ray must undergo at least one reflection before reaching *surface 2*, as opposed to the possibility of direct energy transfer from the source to *surface 1*. This leads to a higher variance in the rays hitting *surface 2*, and thus larger error. A second consequence is that the improvement gained by using a quasi-random sequence is somewhat less pronounced. Table 7.1 summarizes the results of the calculations using the fractional absorptivity method for several values of absorptivity, as well as for a mixed absorptivity case in which the walls are highly reflective ($A = 0.1$), while the wafer is highly absorbing ($A = 0.7$). The mixed case is more typical of an actual reactor. The table gives the convergence rate in terms of β, the expected error for the Halton sequence for $N = 100000$, and the ratio of the expected random error to the Halton error. The expected error after computing for time T^* (defined above) is also given as well as the improvement factor of Halton over random for this case. This table shows that a significant advantage is gained by using the Halton sequence in all cases. Again, the advantage is more prominent for the directly visible *surface 1*.

Figure 7.3 and Table 7.2 provide a similar comparison of the Halton and pseudo-random sequences for the discrete absorption method. As mentioned above, the integrand being evaluated in this method has a larger variance than the integrand associated with the fractional absorption method, and therefore considerably more rays are needed to obtain the same degree of accuracy. However, the computation time required per ray in the discrete case is on average much less. In Figure 7.3 the values of N were chosen so as to have the same computation time as in Figure 7.2, based on the scaling argument given in Section 7.1.2. The dimension of the integral is also much larger, as each reflection corresponds to a separate dimension. Depending on absorptivity, as many as 40 or more dimensions could be required, compared with the three-dimensional integrand of the fractional method (in which the ray path is completely determined by its initial direction). As discussed previously, quasi-random sequences tend to lose their advantage over random sequences as the dimension of the integral increases. This is tempered by the fact that the higher dimensions play a less significant role in the calculation. For example, with an absorptivity of 0.4, 99 per cent of all rays have been absorbed after nine reflections. Nevertheless, this dimensional effect on the quasi-random approach can be seen in Figure 7.3 and Table 7.2. The factor of improvement for Halton over random is noticeably smaller than for the fractional case, although significant gains are still made by using Halton. Again, there is a considerable difference in the accuracy for *surfaces 1* and *2*, most notably in the case of high absorptivity. In this case most rays do not survive the first reflection, so that many fewer rays eventually make it to *surface 2* than *surface 1*.

Table 7.3 summarizes a comparison of the fractional and discrete methods for both a random sequence and the Halton sequence. For the various surfaces and absorptivities the ratio of the expected error using the discrete method to the expected error using the fractional method is shown. In the random case, both methods have the same convergence rate, so this ratio of errors depends only on the variance of the two integrands (computed from the least squares fit of the experimental data) and the computational time scaling factor described above. Thus this value is independent of the com-

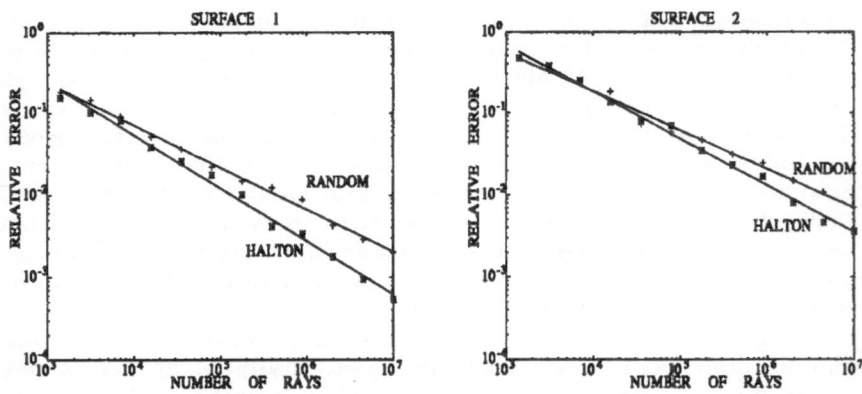

Figure 7.3: Comparison of random and quasi-random sequences using the discrete absorption method with absorptivity = 0.4.

Surface	Abs.	Converg. $\beta : N^\beta$	Error at N $= 8.8 \times 10^5$	Improv. Factor	Error at Time T^*	Improv. Factor
1	0.1	-.54	.0061	1.2	.0061	1.2
	0.4	-.65	.0030	2.4	.0012	2.9
	0.7	-.66	.0020	3.6	.0006	5.1
	mix	-.59	.0020	2.2	.0011	2.5
2	0.1	-.54	.0141	1.2	.0141	1.2
	0.4	-.57	.0141	1.6	.0064	1.8
	0.7	-.57	.0213	1.3	.0071	1.5
	mix	-.57	.0063	2.0	.0036	2.2

Table 7.2: Convergence results for discrete absorptivity method using the Halton sequence.

Surface	Abs.	Random Error Discrete/Fract.	QMC Error Discrete/Fract.
1	0.1	1.70	2.23
	0.4	0.75	1.06
	0.7	0.45	0.71
	mix	0.49	0.65
2	0.1	2.17	2.85
	0.4	0.64	1.76
	0.7	0.96	2.20
	mix	0.49	0.78

Table 7.3: Comparison of error for discrete absorptivity method to error for fractional absorptivity method for random and Halton sequences at T^*.

putation time, or equivalently, of N. However, since the two quasi-Monte Carlo cases do not converge at the same rate, the associated error ratio depends on the computation time. The results are given for time T^*.

At low absorptivities, the fractional method is clearly superior for both random and quasi-random, as well as for both surfaces. As the numbers for the random case indicate, the variance of the discrete method integrand is considerably larger than that of the fractional method. There is a further penalty for the discrete method in the quasi-random case coming from the use of a higher dimensional sequence. As the absorptivity increase, the variance difference becomes small enough to favor the discrete method in the random case. This trend can also be seen for quasi-random for *surface 1*, although the low dimensional advantage of the fractional method moderates the gains seen for the discrete method using the random sequence. *Surface 2*, however, shows rather different behavior, strongly favoring the fractional method for all but the mixed case. This can be explained by noting that when the absorptivity is large, many fewer particles ever make it to the back of the wafer in the discrete case. The "effective" N is then much smaller, so that the quasi-random advantage is considerably less. Figure 7.4 illustrates this by plotting the relative error for the two methods using the Halton sequence for absorptivity 0.7. Results for both *surface 1* and *surface 2* are shown. The abscissa is computation time in arbitrary units, which allows a comparison of the two methods. The scaling was chosen so that one time unit corresponds to the average time needed to compute the path of one ray using the fractional method. On this scale $T^* = 10^5$. In the range of a practical computation, for *surface 1* the advantage of the discrete method

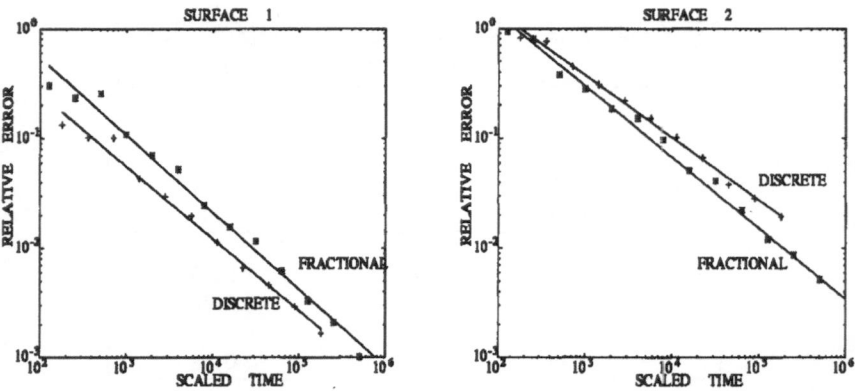

Figure 7.4: Comparison of fractional and discrete absorption methods using the Halton sequence with absorptivity = 0.7.

over the fractional method starts at about a factor of two, but steadily decreases as the accuracy increases. On the other hand, for *surface 2* the fractional method has an increasing advantage over the discrete approach as the computation time increases.

The results for the mixed absorptivity case are shown in Figure 7.5. The success of the discrete method for both surfaces may be explained as follows. The wafer acts as a sink for the rays, absorbing 70 per cent of the rays which strike it. The walls simply reflect most rays. Thus a large percentage of the rays emitted contribute to the statistics of the wafer, both front and back, so that both methods show a comparable improvement over the random case. The relatively low variance of the discrete method in this situation then tilts the balance in favor of discrete over fractional. However, as the graphs show, the advantage of the discrete method again decreases with increasing accuracy. This is a result of the superior convergence rate of the fractional method associated with it being a low dimensional integral. In general, the choice of method may then depend on geometry, the regions of the reactor which are of most interest, and the minimum accuracy required.

To round out the computations, we will briefly describe several other results obtained. A third surface element taken at the edge of the wafer was also observed. The size of this element was 2.36 cm^2, which was much smaller than the other surface elements. Although this element had visibility relative to the heat source similar to that of *surface 1*, the error was almost an order of magnitude larger, and the performance of the Halton sequence was identical or only slightly better than the pseudo-random sequence. This

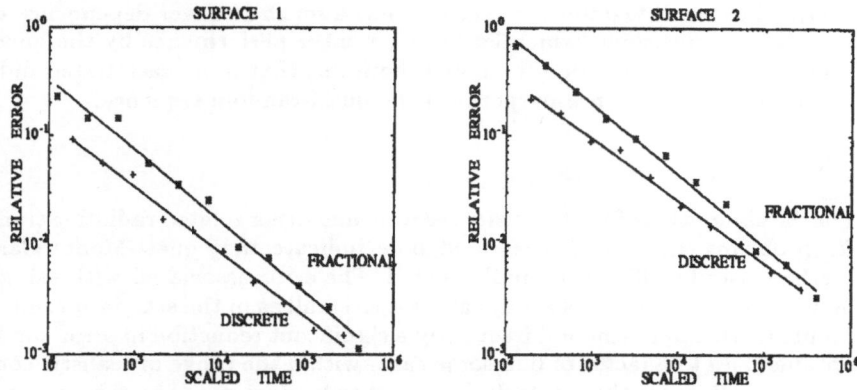

Figure 7.5: Comparison of fractional and discrete absorption methods using the Halton sequence with absorptivity = 0.1 on reactor walls and absorptivity = 0.7 on wafer.

can be attributed to the difference in size of the elements. Many fewer rays strike the smaller surface, leading to an effectively smaller N. Thus for the smaller surface, over the range of N considered, the calculation was still in the range where random and quasi-random are close, similar to what appears in the above graphs at small N. Moreover, the variance of the integrand being evaluated by the simulation is related to the surface size such that the smaller the surface, the higher the variance, and therefore error. This is not surprising, as the use of small surface elements leads to a greater amount of information about the temperature distribution on wafer. To obtain this more refined temperature distribution to the same degree of accuracy of course requires more work (larger N).

Another case considered was that of diffuse reflection at the walls instead of specular. Frequently such cases are often handled by non-Monte Carlo methods. However, as diffuse reflection is easily implemented in a Monte Carlo code, and the results are potentially of interest, a calculation with diffuse reflection for absorptivity 0.4 was carried out. The results using fractional absorption show a significant improvement for Halton over pseudo-random for *surface 1* similar to the results of Figure 7.2. However, for *surface 2*, there results are the same for Halton and random. The difference between the performance for *surfaces 1* and *2* can again be explained in terms of visibility. Much of the energy transferred to *surface 1* comes from rays which have not undergone a reflection. In contrast, to reach *surface 2* requires at least one reflection. Each reflection increases the dimension of the problem by two (two numbers are needed to sample the new direction). Thus the integrand for *surface 2* has a much stronger dependence on the higher dimensions, which leads to a weaker performance by the quasi-random sequence. It should be noted, however, that in no case tested did a pseudo-random sequence outperform the quasi-random sequence.

7.1.4   Conclusions

For the simulation of heat transfer reactors and other related radiative transfer problems, the results presented here indicate that quasi-Monte Carlo methods can be effectively implemented. The error associated with using a quasi-random sequence is always at least as small as in the standard random Monte Carlo approach, and frequently a significant reduction in error can be obtained, up to a factor of 6 in some cases within the range of realistic computations. Because these calculations tend to be dominated by time spent in the ray tracing algorithm, there is no additional computational cost associated with generation of the quasi-random sequence over the pseudo-random sequence. Moreover, the quasi-random sequence may simply be substituted in for the random sequence in existing codes, so that virtually no extra work is required to obtain the quasi-Monte Carlo advantage.

The advantage of quasi-random over random appears to increase as the absorptivity of the reactor increases, for fixed computation time. At lower absorptivities the fractional absorption method is superior to the discrete method. However, as absorptivity increases the discrete method gives better accuracy for surfaces directly visible to the heat source. For the case of highly reflective walls and a highly absorbing wafer, the discrete method was superior for all surfaces, but the advantage over the fractional method decreased with increasing accuracy due to the faster convergence of the fractional method.

The choice of quasi-random sequence (in this case Halton or Sobol') appeared to have little impact on the results. There was also little difference between using a point source or a surface source for the radiation. This additional element of robustness makes quasi-Monte Carlo methods attractive for use in a wide range of radiative transport problems.

7.2   Scalar Control RTCVD-reactor

We now turn to the simulation of a Rapid Thermal Chemical Vapor Deposition (RTCVD) reactor similar to a JIPELEC Jetlight [42][43]. A geometric draft is shown in Figure 7.1. Its water cooled stainless steel walls are axisymmetric and constitute a cold wall system. The shape of the wall builds a reflector to achieve a good temperature uniformity and was designed by simulation [47]. The showerhead inlet is at the top while the illuminator is at bottom. The 6" wafer is upside down such the patterned side is illuminated.

The illumination comes from a lamphouse, which is not axisymmetric by itself, but visible through an axisymmetric opening. There is a single variable control or scalar control of the lamps supported by a pyrometer pointed from the showerhead along the symmetry axis to the wafer center. The lamphouse is separated from the chamber by a thick quartz window with water cooling in between. This water cooling keeps the quartz temperature down although the low conductivity of the quartz could still allow the heating up of the interior of the quartz and constitutes a long time scale transient for the temperature state of the reactor.

For the purpose of simulation, the single variable lamp array was approximated by a uniformly radiating surface at the bottom of the lamphouse. The steady state radiation temperature of the lamps to achieve a wafer temperature around $1040°K$ was assumed to be $2400°K$ with a spectrum of a tungsten halogen lamp. The typical ray path leads from the lamps through the quartz window and the water to the reflector and the wafer. The quartz transmits the optical and the near infrared frequencies of the spectrum (see Figure 5.7 (b)). The water constitutes an optical window which cuts out the near infrared frequencies from the spectrum (see Figure 5.7 (d)). Thus the transmitted part of the spectrum is entirely above the band gap of silicon and is thus absorbed with a temperature independent absorptivity.

For the calculation of the radiation exchange matrix Φ_T a few million rays generated from a Halton number generator were used. For the reactor simulation, PHOENICS/ACCESS-CVD was used. For the determination of the wafer temperature, the atmosphere was taken as N_2 at 5 $Torr$ with a flow rate of 5 slm.

As a first result, the radiation intensity incident on the wafer is shown. The calculation was done for two different reflection models: completely specular and completely diffuse surfaces. The polished stainless steel walls, silicon wafer and quartz walls have a reflection law close to the specular one, but the diffuse model, although physically unrealistic, is contained in many engineering radiation models and serves as a comparison.

Figure 7.6(b) shows the intensity on both sides of the wafer when the wafer has a constant temperature of $300°K$ and the lamps an intensity sufficient to achieve the normal steady state temperature of about $1100°K$. This is a situation similar to the beginning of a temperature ramp. The diffuse and specular model have qualitatively the same illumination characteristic. The showerhead side of the wafer has an increased illumination at the edges due to the reflector and the lamp side illumination in the center. The total illumination shows an increased intensity towards the edge.

The situation is different in Figure 7.6 (a), where the wafer has now a constant temperature of $1100°K$. Here, the total illumination is decreasing towards the edge. This different behavior comes from the self-illumination of the wafer, which represents a non-negligible radiation source at the elevated temperature. The edge losses, which are neglected here corresponding to our

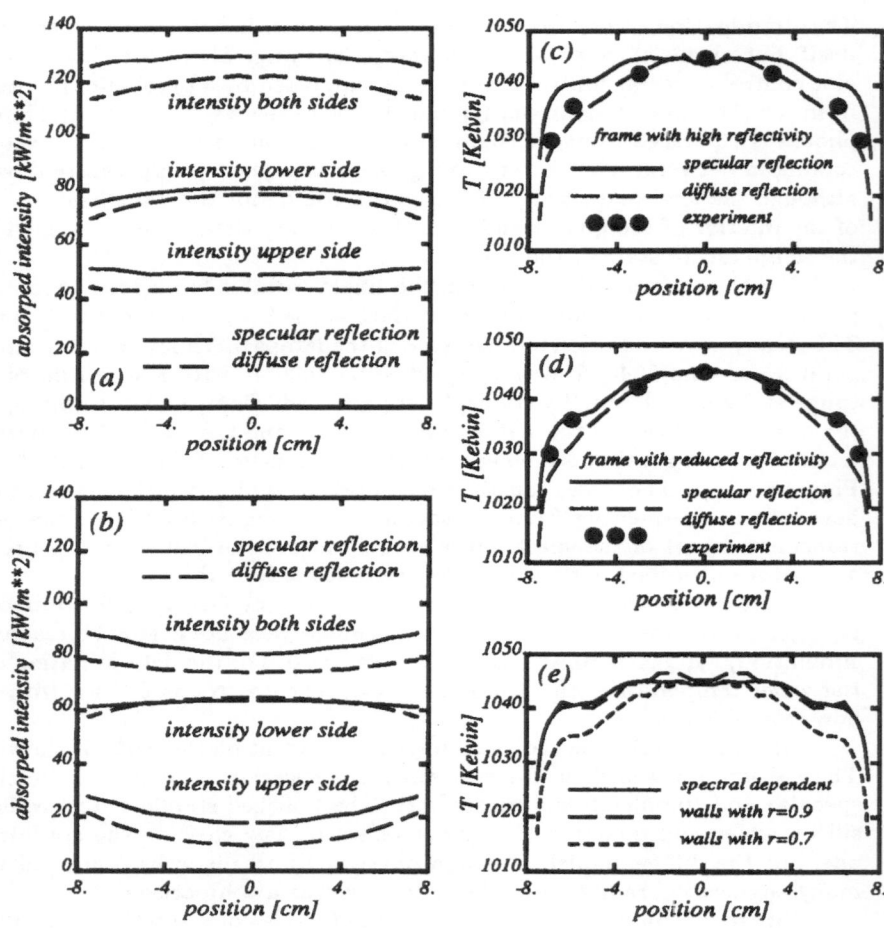

Figure 7.6: Figure (a) and (b) show the illumination intensities of the wafer. In (a), the wafer has a temperature of $1100°K$ (the post-ramp up temperature) and in (b) of $300°K$ (the pre-ramp up temperature). Figures (c), (d) and (e) show the wafer temperature calculated with various models. (c) compares the reflection laws, (d) includes the geometrical details and (e) shows the influence of the spectral dependence.

assumption of a homogeneous wafer temperature, increase this effect if they are present. The presence of the wafer as a further radiation source is also the reason for the higher average intensity in the second case. As a consequence, it is not possible to perform a homogeneous temperature ramp with a single variable radiation source. If the lamps are optimized for steady state, the edges suffer an increased temperature during ramp. This important point was made in the papers of Kakoschke et. al. [39].

The incident intensity is generally lower for the diffuse reflecting surfaces in the case of single sided illumination. This comes from the diffuse reflection law which reflects an equal quantity of radiation forwards and backwards after an interaction with the wall.

Although the intensity profiles on the wafer are qualitatively similar for the diffuse and the specular reflection model, they are quantitatively different on the level of a few percent. This difference is sufficient to result into temperature profiles, which are up to $10°K$ different as is shown in Figure 7.6 (c). For this result, the lamp intensities were adjusted so that the temperatures matched in the center. All spectral optical properties of the surfaces were the same; only the reflection law was different. The temperature profile in the specular case is more homogeneous than for the diffuse case according to the incident intensity profile. The drastic temperature drop at the edge comes mainly from the increased conductive heat loss there. This heat loss results from the large gradient, analogous to the effect of an increased gradient of the electric field near edges [48]. There is a negligible influence of convective cooling at this flow rate. Figure 7.6 (c) also shows some experimental results for comparison. These results are in better agreement with the diffuse model than with the specular model.

This does not, however, prove the specular law to be wrong. There may be certain elements of the model which are not representative of the real reactor and thus introduce other errors. These errors may effectively cancel with the error from the unrealistic diffuse model. In this case, insufficient modeling of the frame region (see Figure 7.1) was assumed to be the main modeling error. The frame was taken to be a polished wall in Figure 7.6 (c). In fact, it is a region with a great deal of geometric detail. When these geometric details are modeled by reducing the frame reflectivity to 0.5, the temperature profile of the specular model fits very well to the experimental data as can be seen in Figure 7.6 (d).

The results just described show the sensitivity of the simulation results to features of the models. Another means of assessing this sensitivity is to compute the heat transfer using a model with constant optical properties instead of spectral dependent optical properties. Figure 7.6 (e) shows the temperature profiles for the specular, spectral model compared to two specular models with a fixed quartz transmissivity of 0.85 and a fixed steel reflectivity of 0.7 and 0.9, respectively. The model with a constant reflectivity of 0.9 is in better agreement with the spectrally dependent model, although the effective reflectivity of steel is closer to 0.7 than to 0.9. The reason lies in the constant transmissivity assumption of the quartz, which is not able to capture the true behavior of the quartz and introduces an error in the calculation. This compensates for the overestimate of the steel reflectivity. This example again shows first order sensitivity of the simulation results to the chosen model, in this case the spectral dependence.

The degree of sensitivity of the simulation which is acceptable is determined by the uniformity requirements of the wafer temperature. In most cases, the temperature may not vary more than a few degrees over the wafer; sometimes the variance must be held to within one degree. As a consequence, a predictive radiation model must contain the correct reflection laws, the correct spectral dependence of the optical properties and sufficient geometrical details to determine the radiation intensity distribution on the wafer to within a few tenths of a percent.

As a final result for this reactor, the temperature of the quartz window on the reactor side can be obtained. Although water cooled, it is raised to $900°K$ in the center.

7.3   Multi-variable Control RTCVD-reactor

As a final example, we present simulation results for a reactor similar to an AG Associates RTCVD reactor. The lamphouse and certain other details, however, were different than that of the original equipment.

The interior of the reactor is axisymmetric, and the illumination comes from the top. The lamphouse itself is not axisymmetric, but consists of nine linear tungsten halogen lamps in rectangular, gold coated cavities inside a cylindrical container. The rectangular cavities serve to improve the control authority of the multi-variable control system. The radiation enters the chamber through a thick, air cooled quartz window. The chamber walls are stainless steel, but the walls are lined with a quartz inlet about 1 mm thick and positioned less than 1 mm away from the walls. The gas flow enters from the top of the inlet through showerhead holes in the chamber. The gas flow leaves the reactor through a long exhaust tube on the bottom. The 8" wafer is rotated to average the asymmetric illumination. To decrease the edge losses, the wafer is surrounded with a guard ring.

The lamphouse region is an example of where the ray tracing is computed with a special subroutine separate from the CFD solver grid (see Section 6.3). Furthermore, several regions of the domain were defined as macroscopic cells to accelerate the ray tracing. Figure 7.7 shows the geometry, grid and special regions of the reactor.

As a first result, the temperatures of the quartz window and quartz inlet were calculated. The heating of the quartz window is a relatively complicated process. The radiation leaves the tungsten filament and travels

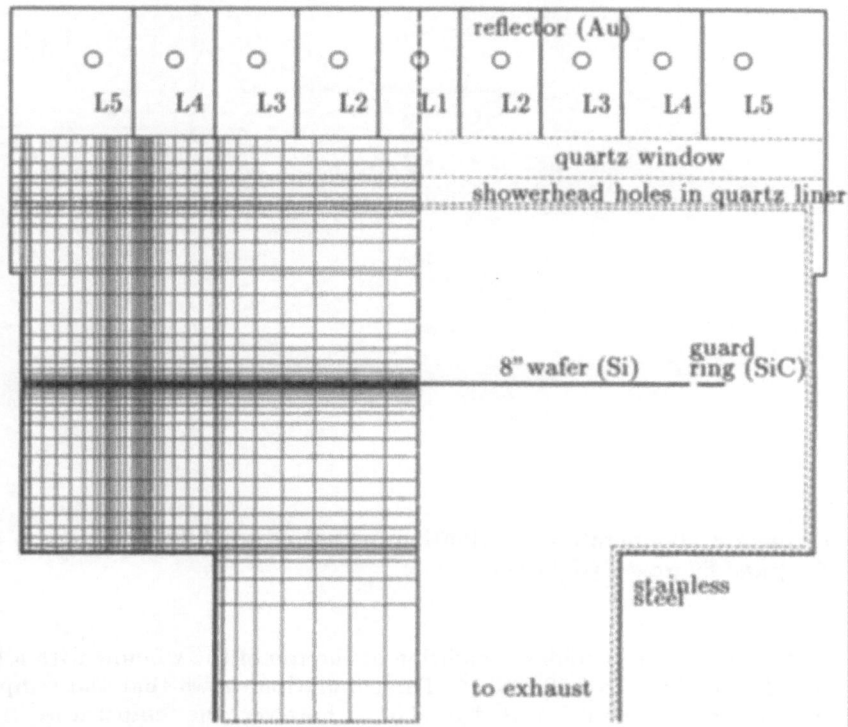

Figure 7.7: Geometry and computational grid of the warm wall reactor.

through a quartz bulb where some of the infrared radiation is absorbed. For the calculation of Φ_T about 10 million rays defined through Halton numbers were used. All surfaces were computed with the spectral dependence shown in Figures 5.5 - 5.7 and assumed to be specularly reflecting.

In the simulation, the bulb was assumed to be a cylinder surrounding the linear filaments $1\,mm$ in radius and $5\,mm$ thick. The quartz bulbs lose some of the absorbed energy conductively to the gold reflector, but only very little radiatively because of the high reflectivity of gold in the infrared. The rest of the energy is transported radiatively and conductively to the window. The major heating, however, is through direct radiation. Because the computational domain is axisymmetric but the lamphouse is not, it is necessary to transfer the energy absorbed in the quartz bulb (a calculation done in the true geometry in the lamphouse subroutine) to axisymmetric solid regions in the computational domain with volume and surface area equivalent to the quartz bulbs. The air cooling of the lamphouse was mod-

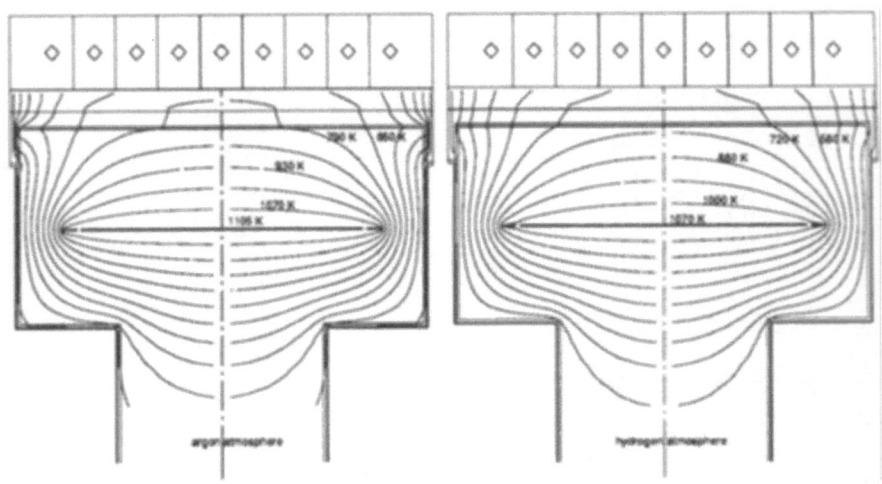

Figure 7.8: Temperature distribution in the reactor filled with argon (left side) and hydrogen (right side).

eled by a Newton boundary condition at the top of the window with a heat transfer coefficient of 10 W/m^2. The simulation shows that the temperature of the glass bulb is raised to $900° K$; however, the temperature of the window depends only slightly on the presence of the bulb.

Behind the quartz window the infrared part of the radiation is lost. Hence the remaining visible and near infrared radiation easily transmits through the showerhead to the wafer. The quartz liner is not heated by the direct radiation. It obtains its heat from the hot wafer mainly radiatively but also conductively. Since the liner is separated from the cooled wall by a gap, it can also heat up. This heat loss from the liner to the wall is now dominated by conduction, such that the final temperature of the liner in steady state depends on the gap width and gas fill. The showerhead obtains its heat from the back-radiation from the hot wafer from below and from the warm window from above. Figure 7.8 shows the temperature distribution in the reactor filled with hydrogen and argon. Convective cooling is neglected. The warm up of window, showerhead and liner introduces temperature drifts of the wafer temperature with different time scales up to several minutes which suggests that a closed loop control should be used.

The wafer temperature may be well controlled with the 5 lamp zones. The controllability can be checked by changing the power setting of the zones. Figure 7.9 shows the wafer temperature for a power setting of 1 kW per lamp plus an additional 1 kW for a single lamp. The uniformity of

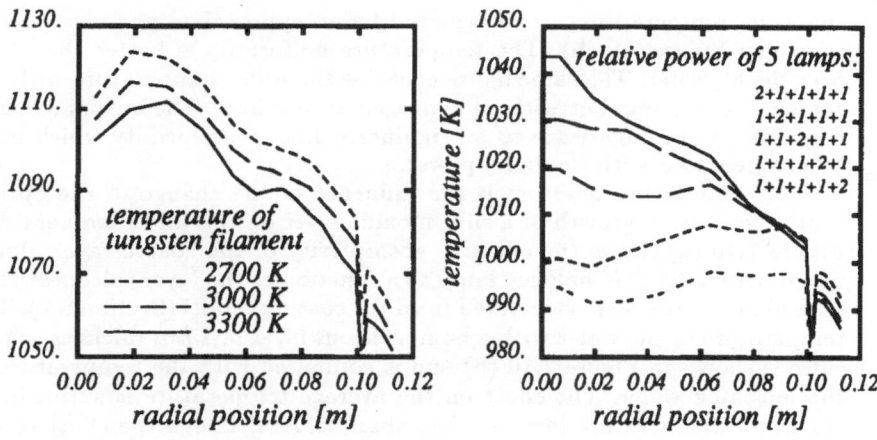

Figure 7.9: The wafer temperature at a fixed power setting depends on the radiation temperature of the tungsten halogen filaments, which is shown on the left hand side. On the right hand side, the controllability of the wafer temperature distribution is demonstrated. A change of the power settings in the lamp arrays gradually changes the wafer temperature distribution.

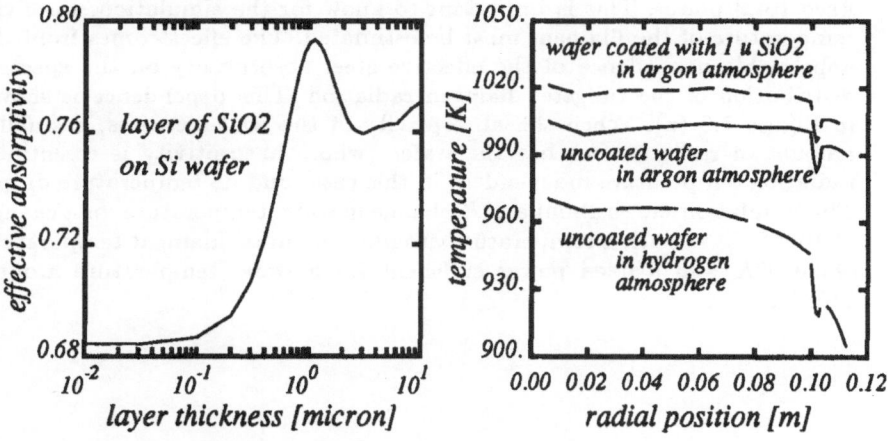

Figure 7.10: On the left hand side the effective absorptivity of a silicon wafer coated with a layer of silicon dioxide SiO_2 is shown. The coherent coating model was used and the spectral absorptivity was averaged with the radiation distribution from a tungsten-halogen lamp at 3300 K. The right hand side shows the effect of the coating on the wafer temperature. It also shows the effect of the kind of gas atmosphere on the wafer temperature. The power setting is fixed in the three cases.

the wafer temperature can be achieved by an appropriate power setting as shown in Figure 7.9 (b). The temperature uniformity is better than $5°K$ over the 8" wafer. The same figure contains the wafer temperature with the same power setting, but with an hydrogen atmosphere. The much increased heat loss at the edge leads to an significant loss of uniformity which must be compensated with the lamp power.

An interesting question is the influence of the change of the optical properties during growth of a silicon oxide layer on the wafer temperature. Figure 7.10 (a) shows the effective absorptivity of the coated layer illuminated with a $3000°K$ halogen lamp as a function of the layer thickness. The optical properties were calculated from the coating model (Section 5.2). The temperature of the wafer with a homogeneous layer of 1 μm thickness on all sides is shown in Figure 7.10 (b) and is compared with the temperature of the uncoated wafer. The effect on the average temperature is surprisingly small: the temperature increases less than $20°K$ although the effective absorptivity increases from 0.7 to 0.8, and the effect on the distribution is negligible. The effect on temperature will be larger in case of a single sided layer and of course in the case of a patterned wafer with a nonhomogeneous coating.

Another effect related to the spectral distribution of the radiation is the dependence of the wafer temperature on the filament temperature with fixed total power. This is important to know for the simulation, since the temperature of the filament must be estimated. The effect comes from the appreciable dependence of the effective steel absorptivity on the spectral distribution of the tungsten halogen radiation. This dependence is shown in Figure 7.9 (a). When the absorptivity of the wall increases, a smaller amount of radiation reaches the wafer (whose absorptivity is essentially radiation temperature independent in this case) and its temperature drops. The simulation shows about a $2°K$ change in wafer temperature for a change of $100°K$ in radiation temperature around a nominal filament temperature of $3000°K$ and a fixed power sufficient for a wafer temperature around $1100°K$.

Chapter 8

Modeling of Charge Transport

Finally we turn to the simulation of charge transport in semiconductor devices. In few other areas has numerical simulation been as successful in aiding design and development. For this reason a great deal of research has been devoted to the modeling and simulation of these devices, and numerous reference texts are available. Because the field is so well covered in the literature, we will not attempt to present a detailed account of device modeling or the wide variety of numerical techniques available. We will instead focus on highlighting certain similarities between rarefied gases and systems of charged particles. For further information on Monte Carlo device simulation we refer the reader to the books of Jacoboni and Lugli [35] and Hess [33]. Information concerning the drift diffusion and hydrodynamic models, as well as other aspects of device simulation, can be found in the books of Markowich, Ringhofer and Schmeiser [60], Hänsch [32], in the series edited by Selberherr, Stippel and Strasser [86], and in the review article by Lee, et al [56].

The fundamental connection is through the Boltzmann equation, which was introduced for both rarefied gases and electrons in semiconductors in Chapter 1. It is a key feature of both systems that they may be described by a distribution function, whether in velocity or in wave number space, which in a very natural limit tends to a Maxwellian distribution. In both cases the importance of the deviations from this equilibrium may be measured in terms of a Knudsen number, which compares the average distance a particle travels between scattering events and a characteristic length of the domain. Figure 2.1 at the beginning of Chapter 2 shows the important transport regimes and associated manufacturing processes for gas dynamics as a function of Knudsen number. Figure 8.1 shows a similar view of the

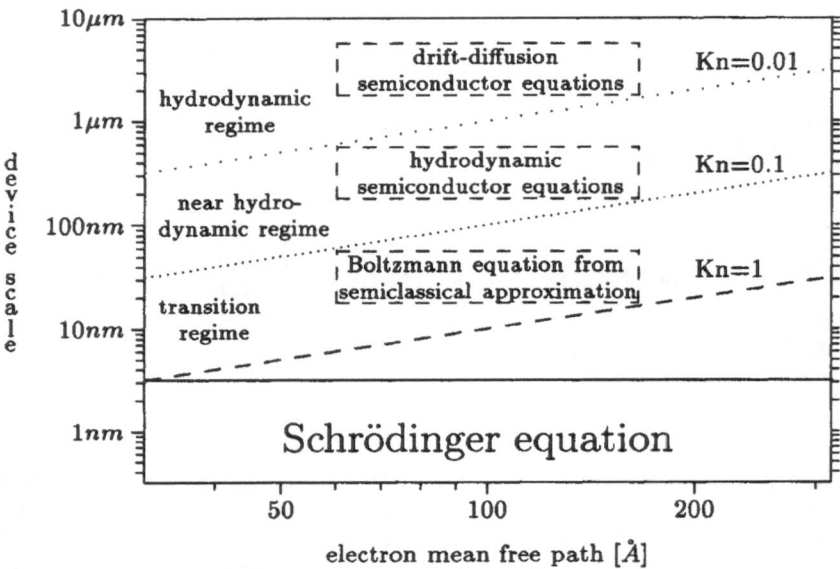

Figure 8.1: Semiconductor equations in relation to the electron mean free path. In contrast to gas molecules, the mean free path is here a complicated function of material and electric field with relatively small variation. When the device scales are comparable to atomic dimensions, the semiclassical approximation breaks down.

various regimes in device simulation. The mean free path λ for electrons is slightly influenced by the application of an external field. For most devices, λ is on the order of 100 Å.

In the regimes where continuum equations are valid, the numerical methods for both gas dynamics and charge transport are similar, and extensive software has been developed in each case. When it is necessary to solve the Boltzmann equation, both areas are again solved by similar methods, in this case Monte Carlo simulation. There is an important difference here, however. For rarefied gases it is usually the case that the non-linear Boltzmann equation must be solved, as the dynamics is determined by interparticle collisions. In semiconductor device modeling, usually electron-electron scattering may be neglected and only scattering off a fixed background need be considered. This leads to a linear Boltzmann equation (which may be coupled non-linearly to a self-consistent field equation). In this sense, the gas dynamics simulation is the more difficult problem. On the other hand, the scattering processes in gas dynamics are relatively straightforward and may be suitably described by simple models such as VHS or

M1 (see Chapter 2). Electron scattering events are more varied and require more detailed modeling than simple elastic collisions. The necessity of computing a self-consistent field also complicates matters. When the linear gas dynamics Boltzmann equation is valid, such as in the case of the sputtering reactor (see Section 4.3.2), there is a significant overlap in the two simulations. Much of the intuition developed by studying the sputtering reactor with the Test Particle method may be applied to charge transport simulations. This is explored in greater depth below. We end by considering two further comparisons of gas dynamics and charge transport, one involving the distribution functions and the other concerning the limiting continuum equations.

8.1 Numerical Methods for Linear Transport

In Section 3.4.1 the Test Particle method for linear transport of a gas species through a fixed background was described. The linearity arises from the assumption that the density of the species of interest is so low that collisions of the particles of this species with themselves do not play a significant role. A further assumption is that the background remains unchanged by the scattering events. Charge transport in semiconductor devices often fulfills these same assumptions. Therefore the numerical methods for the simulation of these two problems share many common aspects.

The Test Particle method of Section 3.4.1 uses a fixed time step to ensure that particles do not travel too far before the possibility of a collision is checked. This approach is more an artifact of the original non-linear method than a necessity. In fact, it is somewhat more natural to sample the time to the next scattering event from an appropriate distribution. This technique is frequently used in charge transport, a detailed account of which may be found in [35]. We present here a summary of the most important aspects.

8.1.1 Event Based Monte Carlo

When the particles are independent of each other and move in an external field, the movement consists of time intervals of free flight followed by collisions. The stochastic calculation of the free flight time followed by the calculation of the collision is called the Event Based Monte Carlo method. During the free flight, the movement in momentum space is defined by Equation (1.116)

$$\hbar \dot{k} = F \tag{8.1}$$

The movement in position space follows with the help of the nontrivial energy dispersion relation and is given by Equation (1.117). The time interval until the next scattering is derived from the collision probability $L(k)$, which contains information on all the different scattering mechanisms. It is assumed that the probability that a scattering event happens in the time interval dt after the time t has the form

$$P(t)i\, dt = L(k(t)) \exp\left(-\int_0^t L(k(t'))dt'\right)\, dt\ . \tag{8.2}$$

The distribution $P(t)$ serves as a probability distribution from which the scattering events can be sampled. The complication in device simulation is that this distribution is different for each particle because of the dependence on k. Therefore methods have been developed to reduce the computational burden. The most prominent is the self scattering method. In this method, the collision probability is assumed to be a constant $\Gamma = 1/\tau$ and is an upper bound for the possible values. The collision probability consists then of the real scattering events $L(k)$ and the self scattering (or dummy) events $L_{self}(k)$, in which no changes occur:

$$\Gamma = L(k) + L_{self}(k). \tag{8.3}$$

The integral equation (8.2) simplifies to the Poisson distribution

$$P(t) = \frac{1}{\tau} \exp\left(-t/\tau\right) \tag{8.4}$$

and the stochastic free flight time t can be easily sampled directly with the help of a random number R_t as

$$t = -\tau \ln(1 - R_t). \tag{8.5}$$

A variety of methods exist to estimate Γ. It is important that the estimate not be too large to avoid excessive computation of self scattering events. One common approach is to construct Γ as a step function which is constant on various subintervals of the electron energy domain. If the electron remains within a given energy subinterval during its free flight, then the Γ for that subinterval should provide a close upper bound for the scattering probability so that relatively few self scattering events are necessary. If the electron crosses into another energy subinterval, care must be taken that an appropriate combination of the two Γ's is used.

The event based Monte Carlo technique may be applied to all linear transport problems with scattering in the volume. This includes of course charge transport in semiconductors and trace species transport through a fixed background gas. These methods also apply to neutron transport and radiation transport through an absorbing atmosphere.

8.1.2   History Based Monte Carlo

When the particles are not independent of each other and move in a self-consistent field, as is actually the case in a semiconductor, a different Monte Carlo method should be chosen. According to Equation (1.123), the calculation of the effective electric field depends on the electron number density $n(x, t)$. This number density may be calculated in a Monte Carlo computation very similar to the gas dynamics case by constructing a cell structure in the domain and sampling the particles at appropriate time steps. The time step has to be rather small in device simulation, about $10^{-14} s$, to avoid spurious plasma oscillations. When a whole ensemble of particles is simulated simultaneously (Ensemble Monte Carlo method), this leads to a number density distribution for every time step. Problems arise now with the event based Monte Carlo, however, because the individual scattering event times do not correspond to the ensemble time step. This leads to the History Based Monte Carlo method, in which a common time step for all particles is prescribed. This time step must be rather small so that the expected number of collision per particle is most likely smaller than one. Consequently, not all particles undergo a scattering event in this common time step. This method is then very close to the Test Particle method of Section 3.4.1 The history based Monte Carlo method is very well suited for vectorization; therefore it is the method of choice in modern Monte Carlo device simulators [81].

8.2   Further Comparisons

We now consider two further areas in which a comparison of gas and charge transport is of interest. The first concerns the nature of the particle distribution function as the Knudsen number changes from the equilibrium, collision dominated regime to the transition regime. The second involves the mass and charge current equations which are valid in the equilibrium limit.

8.2.1   Distribution Functions

In the small Knudsen number regime the mean free path of a particle is so small compared to the macroscopic length scale (or equivalently the relative macroscopic gradients of the flow) that particle collisions and scattering events dominate. The result is a Maxwellian distribution function for the particle velocity of a gas molecule or the wave vector of an electron (in the parabolic band structure approximation, these are equivalent). For gas dynamics, the ansatz of a Maxwellian distribution leads to the Euler equations. For electrons, if the temperature of the distribution is taken to be the lattice temperature, the resulting macroscopic equations are the drift diffusion equations.

As the macroscopic length scales decrease (or the gradients increase), the Maxwellian approximation begins to break down. The first order corrections (in Knudsen number) from the Chapman-Engskog expansion for gas dynamics leads to the viscosity and heat conduction terms of the Navier-Stokes equations. For electron transport, the next approximation is the hydrodynamic equations. These again usually assume a Maxwellian distribution, but now the electron temperature is treated as a variable. The fact that energy is not conserved by the electrons scattering off the background lattice may lead to a rather complicated energy equation.

Practically, there is an important difference between gas dynamics and electron transport as the Knudsen number becomes larger which has not as yet been addressed. This concerns the high energy tail of the distribution function. For a non-reacting gas, the nature of this tail is generally not important, except perhaps if a high energy source is present (as in a sputtering reactor). The important transport information is contained in the portion of the distribution function near the stream velocity. The same is true for the bulk flow of electrons, the majority of which are distributed very close to Maxwellian with the lattice temperature. However, electrons in the high energy tail, as opposed to the high energy gas molecules, often may be connected to important physical phenomena. An example of this is device damage. This occurs when high energy electrons penetrate the insulating oxide regions of a device. Over time this can result in a change in the electrical properties of the device. Thus it is desirable to have detailed knowledge of the high energy tail to help predict the degree of damage which may occur.

Device damage is analogous to chemical reactions in a plasma. In this case, the vast majority of particles do not have high enough energy to overcome the activation energy of the reaction. It is only those particles in the high energy tail which do. These reactions, although rare, may play a dominant role in the behavior of the system. Thus in this case knowledge of the tail is also vital.

The classical semiconductor equations do not provide information about the true nature of the high energy tail. This can be obtained only from solving the Boltzmann equation with a Monte Carlo method. This method, however, is so computationally expensive, that at this time it cannot be realistically considered as a simulation tool. However, the Monte Carlo method may be used to check the validity of proposed analytic models of the high energy tail. A number of such models have been proposed in the literature [32]. We present here a brief comparison of two common models with a Monte Carlo calculation following the work of Vogelsang [95].

The first ansatz for the distribution function tail is the Lucky-Electron model. This model assumes that the high energy electrons are produced by

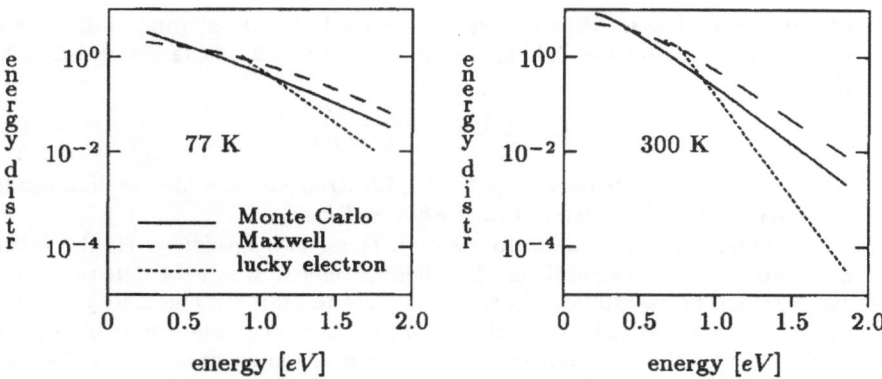

Figure 8.2: Comparison of Lucky-Electron and Maxwellian models and a Monte Carlo simulation of the Boltzmann equation for the high energy tail of the electron energy distribution function. The plot on the left is at $77°K$, while the plot on the right is at $300°K$.

acceleration due to the electric field while in free flow. The longer the free flight before a collision, the higher the energy. The distribution function for the high energy electrons therefore depends strongly on the mean free path λ. Because the probability of traveling a distance greater than a mean free path falls off exponentially with the distance, the distribution function is assumed to have the form

$$f_{LE}(\epsilon) \propto \exp\left(-\frac{\epsilon}{qE\lambda}\right) . \qquad (8.6)$$

Here ϵ is the electron energy, q is the charge and E is the field strength. For transport in silicon, typical values of these parameters are $E = 10^5 \ V/cm$ and $\lambda = 100 \mathring{A}$, so that $qE\lambda \approx 100 \ meV$. It should be noted that this distribution function is only valid for energies ϵ well above the mean electron energy.

At the other extreme is the Maxwellian model for which it is assumed that electron-electron collisions dominate so as to bring the system into thermodynamic equilibrium at some electron temperature T_e (higher than the lattice temperature due to the acceleration from the electric field). In terms of the wave vector k, the Maxwellian distribution is expressed as

$$f_{MW}(k) \propto \exp\left(-\frac{(\hbar k - mv_d)^2}{2mk_B T_e}\right) \qquad (8.7)$$

where v_d is the electron drift velocity. In the parabolic effective mass approximation, for typical drift velocities in silicon and for energies above 1

eV, it can be shown that the drift velocity has only a minor influence on the energy distribution. Thus, the high energy tail of the distribution has the form

$$f_{MW}(\epsilon) \propto \exp\left(-\frac{\epsilon}{k_B T_e}\right) . \tag{8.8}$$

The difference then between the Lucky-Electron and the Maxwellian models lies only in the difference between $qE\lambda$ and $k_B T_e$.

Figure 8.2 shows a comparison of these two models with the results of a Monte Carlo calculation. The field strength was held constant at $3 \cdot 10^5$ V/cm. The results are shown at $77^\circ K$ and $300^\circ K$. The Lucky-Electron model was normalized to match the Monte Carlo calculation at 0.9 eV and $300^\circ K$, while the Maxwell model was normalized so that it had the same mean energy as the distribution function calculated in the Monte Carlo simulation. The results show that at high energies, the true distribution function lies between the two extremes.

8.2.2   Mass and Charge Currents

The hydrodynamic equations for gas flow from Section 2.2 express the mass diffusion current of a species at constant temperature as a linear combination of a density gradient and an external force. For a binary mixture (assuming equal molecular masses), upon inclusion of external forces, Equation (2.27) may be written

$$\vec{j}_1 = m D^{mass}\left(\nabla n_2 - \frac{n_2}{k_B T}F_2\right) - m D^{mass}\left(\nabla n_1 - \frac{n_1}{k_B T}F_1\right) \tag{8.9}$$

where D^{mass} is the mass diffusion coefficient, n_i is the number density of species i and F_i is the force acting on species i. The charge current equation (1.146) from the drift diffusion system shows a similar structure whereby the charge current is expressed as a linear combination of a density gradient and a force. With the help of the Einstein relation (Equation (1.147)), this may be written

$$J = q D^{charge}\left(\nabla n - \frac{n}{k_B T}F\right) . \tag{8.10}$$

Here D^{charge} is the charge diffusivity and $F = -qE_{eff}$ is the force on an electron. The fact that the gradient and the force term may be united by a single diffusion coefficient in both the gas and electron case is notable. This is a consequence of both systems being close to thermodynamic equilibrium. As nonequilibrium effects arise with increasing Knudsen number, the Einstein relation

$$D^{charge} = \frac{k_B T}{q}\mu \tag{8.11}$$

relating the electron mobility μ in an external field to the electron diffusion coefficient breaks down. A similar phenomenon occurs in gas dynamics.

The main effect of the nonequilibrium character of the flow - other than the hot electron effects of the high energy tail - is the modification of the transport coefficients D^{charge} and μ. In order to extend the validity of the drift diffusion equations to the higher Knudsen number regime, hybrid schemes between the Monte Carlo simulation and the drift diffusion equations have been introduced [9]. In this approach a Monte Carlo simulation is used in critical regions to provide flux corrections to the charge current given in Equation (1.146). These flux corrections can be compared with the mass diffusion flux corrections discussed in Section 4.1.2. The charge current corrections are obtained by computing the charge diffusivity and electron mobility as moments of the non-equilibrium distribution function computed in the Monte Carlo calculation. The corresponding expressions can be found in [9] and in [95] in a generalized form including a thermal diffusion term.

Appendix A

Monte Carlo Methods

The fundamentals of Monte Carlo integration are presented here. As actual applications of Monte Carlo tend to be very problem specific, only the most basic technique is described in detail. Much more complete accounts may be found in the books of Hammersley and Handscomb [31] and Kalos and Whitlock [40]. Some general comments are then offered on applications and modifications. Finally, questions of practical implementation are addressed including random number generation and stopping criteria.

The basics of the Monte Carlo method are rooted in probability theory [28]. For our purposes it is sufficient to consider a probability space Ω which is a subset of $\mathbf{R}^n$ equipped with a probability measure $d\mu$ which can be expressed as

$$d\mu = f(x)\, dx \ (x \in \Omega) \, . \tag{A.1}$$

Here dx is Lebesgue measure and $f(x)$ is some non-negative real-valued integrable function such that

$$\int_\Omega f(x)\, dx = 1 \, . \tag{A.2}$$

A random variable χ on Ω is then defined through an integrable function (with respect to $d\mu$)

$$\chi : \Omega \to \mathbf{R} \, . \tag{A.3}$$

The value of χ is random in $\mathbf{R}$, with the probability that χ lies within the real interval (a, b) defined by the formula

$$P(a < \chi < b) = \int_A d\mu \tag{A.4}$$

where

$$A = \{x \in \Omega \ : a < \chi(x) < b\} \, .$$

219

The expected value of a random variable χ is defined as

$$E(\chi) = \int_\Omega \chi(x)\, f(x)\, dx \; . \tag{A.5}$$

Two random variables ϕ and ψ are said to be identically distributed if

$$P(a < \phi < b) = P(a < \psi < b) \tag{A.6}$$

for all intervals (a, b). This corresponds to them having identical generating functions, i.e. $\phi(x) = \psi(x)$ a.e.. Two random variables ϕ and ψ are said to be independent if

$$P(a < \phi < b \text{ and } c < \psi < d) = P(a < \phi < b) \cdot P(c < \psi < d) \; . \tag{A.7}$$

One basic property of independent random variables ϕ and ψ is that

$$E(\phi \cdot \psi) = E(\phi)\, E(\psi) \tag{A.8}$$

The concept of a random variable may be extended to several dimensions by allowing the function $\chi(x)$ to be vector valued, mapping R^n to R^m. In particular, if $m = n$ and $\chi(x) = x$, then the random vector x is said to have probability density function (pdf) $f(x)$. Using the generalization of the above definition for the probability of a random variable being in an interval, it follows that the probability of the random vector x being in a set $A \subset \Omega$ is just $\int_A d\mu$. Moreover, any random variable can be thought of as being a real-valued function of a random vector x with pdf $f(x)$. Conversely, any integrable real-valued function of a random vector x with pdf $f(x)$ is a random variable.

The Monte Carlo numerical integration method is based on the following observation. Suppose that $g(x)$ is an integrable function on Ω with respect to $d\mu$, and we are interested in evaluating

$$I = \int_\Omega g(x)\, f(x)\, dx \; . \tag{A.9}$$

Then the value of this integral is just the expectation of the random variable g. If $x_i, i = 1, \ldots, N$ is a sequences of independent, identically distributed random variables with pdf $f(x)$, then the quantity

$$G_N = \frac{1}{N} \sum_{i=1}^{N} g(x_i) \tag{A.10}$$

is also a random variable. G_N serves as an estimate of the true expectation of $g(x)$. Because the x_i are identically distributed, it follows that $I = E(g) = E(g(x_i))$ for all i. Consider the function

$$\epsilon(x) = I - G_N(x) \; . \tag{A.11}$$

If $\epsilon^2(x)$ is integrable (with respect to $d\mu$), then ϵ^2 is also a random variable, and we can calculate its expectation. Using the fact that I is constant, so that $E(I) = I$, and the independence of the x_i, we find that

$$
\begin{aligned}
E(\epsilon^2) &= E(I^2) - 2E(I \cdot G_N) + E(G_N^2) & \text{(A.12)} \\
&= I^2 - 2I \left(\frac{1}{N} \sum_{i=1}^{N} E(g(x_i)) \right) \\
&\quad + \frac{1}{N^2} \sum_{i=1}^{N} E(g(x_i)^2) + 2\frac{1}{N^2} \sum_{i=1}^{N-1} \sum_{j=i+1}^{N} E(g(x_i)g(x_j)) \\
&= -I^2 + \frac{1}{N} E(g^2) + (1 - \frac{1}{N})I^2 \\
&= \frac{E(g^2) - I^2}{N} \\
&= \frac{\sigma^2}{N}
\end{aligned}
$$

Here $\sigma^2(g)$ is the variance of the random variable g, defined as

$$ \sigma^2 = E(g^2) - (E(g))^2 . \quad \text{(A.13)} $$

The quantity σ (the square root of the variance) is known as the standard deviation. It follows then that the expectation of ϵ^2 will exist if the random variable g has finite variance, i.e. the function $g(x)$ is in $L_2(\Omega)$.

The quantity ϵ is interpreted as the error in using G_N to approximate I (the expectation of g). Practically, this means that the integral of $g(x)$ (w.r.t. $f(x)dx$) is approximated by averaging the value of $g(x)$ at N random points x_i in the domain Ω, where the x_i are distributed according to $f(x)$. Thus Monte Carlo integration is simply a kind of quadrature where the integration nodes are random points and each node has weight $1/N$. An important distinction of Monte Carlo is that the error ϵ is also random, so statements concerning convergence can only be given in terms of expectation and probability.

Equation (A.12) shows that the expectation of the square of the error decreases like $1/N$ (it is important to note that the variance $\sigma^2(g)$ is independent of N). Thus the basic Monte Carlo convergence rate is

$$ \epsilon \approx \frac{\sigma(g)}{\sqrt{N}} . \quad \text{(A.14)} $$

For any given realization of Equation (A.10) using N random points, the actual error may be quite different from the expected value. Also, although the expected error converges like $1/\sqrt{N}$, adding in more random points to a given calculation (that is, increasing N) does not guarantee that the error will decrease. However, it is possible to estimate the probability that the error will be smaller than a given value by using the Central Limit Theorem. As applied to this case, this theorem states that

$$\lim_{N \to \infty} P\left(|\epsilon| < \frac{\sigma}{\sqrt{N}} \delta\right) = \sqrt{\frac{2}{\pi}} \int_0^{\delta} \exp[-t^2/2] \, dt \; . \tag{A.15}$$

Informally, this may be interpreted as saying that as $N \to \infty$, the probability density function associated with the random variable G_N is a Gaussian distribution with mean $E(g)$ and variance $\sigma^2(g)/N$. The size of N necessary before this asymptotic result can be observed depends of course on $g(x)$, but frequently it is assumed to hold, independent of N.

As an example, consider the integral

$$\int_0^1 x^2 \, dx = 1/3 \; . \tag{A.16}$$

We may consider sampling random numbers from a uniform distribution on $[0, 1]$ corresponding to $f(x) = 1$. Then $g(x) = x^2$, and the Monte Carlo approximation to the integral is

$$G_N = \frac{1}{N} \sum_{i=1}^{N} x_i^2 \tag{A.17}$$

where the x_i are uniform random numbers on $[0, 1]$. The variance associated with this calculation is

$$\sigma^2(g) = \int_0^1 x^4 \, dx - (1/3)^2 = 4/45 \; . \tag{A.18}$$

An alternative approach would be to set $f(x) = 2x$ and $g(x) = x/2$. Then if $\{\tilde{x}_i\}$ is a sequence of N random numbers with pdf $f(x)$, the integral is approximated by

$$G_N = \frac{1}{N} \sum_{i=1}^{N} \tilde{x}_i/2 \; . \tag{A.19}$$

In this case, the variance of the integrand is

$$\sigma^2(g) = \int_0^1 (x/2)^2 (2x) \, dx - (1/3)^2 = 1/72 \; . \tag{A.20}$$

Thus the variance, and therefore the error, will be considerably lower if the second method is used. This illustrates an important aspect of Monte Carlo: the size of the error may be reduced through a skillful choice of the probability density function for the random nodes. The fundamental tasks of the method are then choosing a suitable pdf $f(x)$ and generating a sequence of random points which have $f(x)$ as their pdf. A wide variety of variance reduction techniques, based on both physical and mathematical reasoning, have been developed for different classes of problems. The approach illustrated in this example is known as importance sampling.

Because of the slow convergence rate and statistical uncertainty of the results, Monte Carlo is generally not used for low dimensional integrals of smooth functions which are better handled by grid based quadrature methods. As dimension increases, however, the convergence rate of grid methods decays and the number of grid points required grows exponentially. Whether Monte Carlo or a quadrature rule is better for problems in less than 10 dimensions depends a great deal on the integrand. Above ten dimensions a grid based method is not feasible.

A typical application of Monte Carlo integration is to evaluate an integral of the form

$$\int_\Omega g(x)\, F(x)\, dx \tag{A.21}$$

where the domain Ω is fairly high dimensional and the function $F(x)$ is a probability density function known only as

$$F(x) = \frac{f(x)}{\int_\Omega f\, dx} \tag{A.22}$$

for a given $f(x) \geq 0$. Sophisticated sampling techniques have been developed for obtaining random points x distributed according to $F(x)$ without explicitly calculating the normalization factor $\int_\Omega f\, dx$.

Often the function $f(x)$ is itself not explicitly known, but given only as the solution of an integral equation of the form

$$f(x) = S(x) + \int_\Omega K(x|y)\, f(y)\, dy . \tag{A.23}$$

Here $K(x|y)$ is a transition kernel describing the probability of starting at point y in phase space Ω and scattering to x. $S(x)$ is a source term. Radiative and neutron transport problems are often formulated in this manner. Points are sampled from $f(x)$ by performing a particle simulation, whereby N particles, initially sampled from $S(x)$, are traced through a series of scattering events. Monte Carlo techniques are also used to sample from density functions which are described by more complicated integral equations, such as the time dependent, non-linear Boltzmann equation.

As a practical matter, the random numbers used in Monte Carlo computations are rarely actually random. Some attempts have been made to harness physical random processes as a source, but these suffer from the difficulty of ensuring that the underlying distribution remains constant in time. Occasionally tables of reliable random numbers stored on disks are used. However, by far the most common sources of random numbers are deterministic pseudo-random generators. These are deterministic algorithms which map the sequence of integers $1, 2, \ldots$ to a sequence of real numbers between 0 and 1 (or possibly a sequence of real vectors in the d-dimensional unit cube). The goal of a good pseudo-random generator to produce a sequence which has many of the same properties as a uniformly distributed random sequence (or more precisely, the same properties as a sequence of independent, identically distributed random variables with probability density function $f(x) = 1$ on the domain $\Omega = (0, 1)$). Once a uniform random (or pseudo-random) sequence is available, various techniques may be applied to map that sequence to a sequence with the desired distribution $f(x)$. No deterministic sequence can match all the properties of a random sequence. However, depending on the application, some properties are more important than others for the success of the Monte Carlo method. A battery of statistical tests exists for evaluating the quality of a pseudo-random sequence. These are useful in discovering correlations among the terms of the sequence which may have a detrimental effect in a Monte Carlo calculation. See [75] for more details on random and pseudo-random sequences.

Finally, the question of how large an N to use should be considered. Consider the Monte Carlo evaluation of the integral $E(g) = \int g(x) f(x) dx$. If the variance of g is known and the Central Limit Theorem is assumed to hold, then a value of N may be determined that will guarantee with a certain probability that the error is smaller than a given limit. However, in practise, if $E(g)$ is unknown, it is unlikely that $\sigma^2 = E(g^2) - (E(g))^2$ will be known. Of course this may also be estimated during the Monte Carlo calculation; however the variance of the variance (related to $E(g^4)$) will often be quite large relative to the variance. Thus, one must be careful in determining a stopping criterion and with the statement of the certainty with which the answer is correct.

Bibliography

[1] Fidap. Technical report, Fluid Dynamics Int., Evanston, Il, USA.

[2] Fluent. Technical report, Fluent Inc., Lebanon, NH, USA.

[3] Phoenics/Access-CVD. Technical report, Cham Ltd., Wimbledon, London, Great Britain.

[4] ARS Software, Landover, MD 20785. *OPTIMATR, A Computer Program to Calculate Optical Properties of Materials*, 1993.

[5] D. E. Aspnes and A. A. Studna. *Applied Optics*, 14:220, 1975.

[6] R. M. A. Azzam and N. M. Bashara. *Ellipsometry and Polarized Light*. North Holland, Amsterdam, 1977.

[7] H. Babovsky. On a simulation scheme for the Boltzmann equation. *Mathematical Methods in the Applied Sciences*, 8:223–233, 1986.

[8] H. Babovsky et al. Application of well distributed sequences to the numerical simulation of the Boltzmann equation. *J. Comp. Appl. Math.*, 31(1):15–22, 1990.

[9] S. Bandyopadhyay, M. E. Klausmeier-Brown, C. M. Maziar, S. Datta, and M. S. Lundstrom. A rigorous technique to couple Monte Carlo and drift diffusion models for computationally efficient device simulation. *IEEE Trans. on Electron Devices*, ED-34:392, 1987.

[10] J. J. M. Beenakker, F. R. McCourt, W. E. Köhler, and I. Kuščer. *Nonequilibrium Phenomena in Polyatomic Gases*. Clarendon Press, Oxford, 1990.

[11] G.A. Bird. *Molecular Gas Dynamics*. Clarendon Press, Oxford, 1976.

[12] G.A. Bird. Simulation of multi-dimensional and chemically reacting gas flows. In R. Campargue, editor, *Rarefied Gas Dynamics*, pages 365–388. CEA, Paris, 1979.

[13] G.A. Bird. Monte Carlo simulation in an engineering context. *Prog. Astro. Aero.*, 74:239–255, 1981.

[14] G.A. Bird. Definition of mean free path for real gases. *Physics of Fluids*, 26(11):3222–3, 1983.

[15] C. Borgers, C. Greengard, and E. Thomann. The diffusion limit of free molecular flow in thin plane channels. *SIAM J. Appl. Math.*, 52(4):1057–75, 1992.

[16] M. Born and E. Wolf, editors. *Principles of Optics*. The Macmillan Company, New York, 2d rev. ed. edition, 1964.

[17] P. Bratley and B.L. Fox. Implementing Sobol's quasirandom sequence generator. *ACM Trans. Math. Software*, 14:88–100, 1988.

[18] M. Q. Brewster. *Thermal Radiation Transfer and Properties*. John Wiley, 1992.

[19] R. P. Brinkmann, K. Hsiau, J. Zheng, and J. P. McVittie et al. Ion-heating in the pre-sheath of an capacitively coupled rf-discharge. In *Proc. of the Tenth Symp. on Plasma Proc., San Francisco, 1994*, 1994.

[20] C. Cercignani. *The Boltzmann Equation and its Applications*. Springer-Verlag (Applied Mathematical Sciences, v. 67), New York, 1988.

[21] S. Chapman and T.G. Cowling. *The Mathematical Theory of Non-uniform Gases*. Cambridge University Press, 1939.

[22] F. Coron. Derivation of slip boundary conditions for the Navier-Stokes system from the Boltzmann equation. *Journal of Statistical Physics*, 54(3/4):829, 1989.

[23] D. G. Coronell. *Simulation and Analysis of Rarefied Gas Flows in Chemical Vapor Deposition Processes*. PhD thesis, Massachusetts Institute of Technology, Cambridge, 1993.

[24] D. G. Coronell and K. F. Jensen. Analysis of transition regime flows in low pressure chemical vapor deposition reactors using the direct simulation Monte Carlo method. *J. Electrochem. Soc.*, 139(8):2264, 1992.

[25] S. R. de Groot and P. Mazur. *Non-Equilibrium Thermodynamics*. North-Holland, Amsterdam, 1962.

[26] A. DeSanto and G.S. Brown. Analytical techniques for multiple scattering from rough surfaces. In E. Wolf, editor, *Progress in Optics Vol. XXIII*. North Holland, Amsterdam, 1986.

[27] H. Faure. Discrépance de suites associées à un système de numération (en dimension s). *Acta Arithmetica*, 41:337–351, 1982.

[28] W. Feller. *An Introduction to Probability Theory and Its Applications, Volumes I and II*. John Wiley and Sons, 1966.

[29] H. Grad. Principles of the kinetic theory of gases. In *Handbuch der Physik Vol.XII*. Springer Verlag, 1958.

[30] E. L. Hall. Rapid thermal processing: National, Sematech and SRC Roadmaps. In B. Lojek, editor, *Proceedings of the 1st International Rapid Thermal Processing Conference*, page 22, 1993.

[31] J.M. Hammersley and D.C. Handscomb. *Monte Carlo Methods*. Methuen, London, 1964.

[32] W. Hänsch. *The Drift Diffusion Equation and Its Applications in MOSFET Modeling*. Springer-Verlag, Wien, 1991.

[33] K. Hess, editor. *Monte Carlo Device Simulation: Full Band and Beyond*. Kluwer Academic Publishers, Boston, 1991.

[34] J. O. Hirschfelder, C. F. Curtiss, and R. B. Bird. *Molecular Theory of Gases and Liquids*. John Wiley, New York, 1954.

[35] C. Jacoboni and P. Lugli. *The Monte Carlo Method for Semiconductor Device Simulation*. Springer-Velag, Wien, 1989.

[36] K. F. Jensen. Fundamentals of chemical vapor deposition. In M. L. Hitchmann and K. F. Jensen, editors, *Chemical Vapor Deposition - Principles and Applications*. Academic Press, London, 1993.

[37] J. T. Kajiya. Anisotropic reflection models. *Computer Graphics*, 19(3):15, 1985.

[38] J. T. Kajiya. The rendering equation. *Computer Graphics*, 20(4):143, 1986.

[39] R. Kakoschke, E. Bußmann, and H. Föll. The appearance of spatially nonuniform temperature distributions. *Appl. Phys. A*, 52:42, 1992.

[40] M.H. Kalos and P.A. Whitlock. *Monte Carlo Methods, Volume I*. John Wiley and Sons, 1986.

[41] R. J. Kee, F. M. Rupley, and J. A. Miller. Chemkin II: A Fortran chemical kinetics package for the analysis of gas-phase chemical kinetics. Technical report, SANDIA National Laboratories report SAND89-8009B.UC-706, Albuquerque, 1989.

[42] A. Kersch and W. Morokoff. Accelerated Monte Carlo modeling of an RTCVD-reactor. *IEDM 93 Technical Digest*, page 869, 1994.

[43] A. Kersch and W. Morokoff. Radiative heat transfer with Quasi Monte Carlo methods. In E. Strasser S. Selberherr, H. Stippel, editor, *Simulation of Semiconductor Devices and Processes, Vol.5*, page 373. Springer Verlag, 1994.

[44] A. Kersch, W. Morokoff, and Chr. Werner. Selfconsistent simulation of sputter deposition with the Monte Carlo method. *Journ. Appl. Phys.*, 75(4):2278, 1994.

[45] A. Kersch, W. Morokoff, Chr. Werner, D. Restaino, and B. Vollmer. Modeling of a sputter reactor using the direct simulation Monte Carlo method. *IEDM 92 Technical Digest*, page 181, 1993.

[46] A. Kersch, W.J. Morokoff, and A. Schuster. Radiative heat transfer with quasi-Monte Carlo methods. *Transport Theory and Stat. Phys.*, 23(7), 1994. Marcel Dekker, Inc., NY.

[47] A. Kersch, H. Schaefer, and Chr. Werner. Improvement of thermal uniformity of rtp-cvd equipment by application of simulation. *IEDM 91 Technical Digest*, page 883, 1992.

[48] A. Kersch and Th. Schafbauer. Modeling of convective cooling at the wafer edge. *IEEE Trans. on Semicond. Manuf.*, 1994. to appear.

[49] Ch. R. Kleijn. Chemical vapor deposition processes. In M. Meyyappan, editor, *Computational Modeling in Materials Processing*. Artech House, Boston, 1994.

[50] Chris R. Kleijn and Chr. Werner. *Modeling of Chemical Vapor Deposition of Tungsten Films*. Birkhäuser Verlag, Basel, 1993.

[51] K. Koura. Null-collision technique in the direct-simulation Monte Carlo method. *Physics of Fluids*, 29:3509–3511, 1986.

[52] M. Krook and T.T. Wu. Exact solutions of the Boltzmann equation. *Physics of Fluids*, 20:1589–1595, 1977.

[53] E. P. Lafortune and Y. D. Willems. A theoretical framework for physically based rendering. *Computer Graphics Forum*, 13(2):97, 1994.

[54] P. L. Landau and E. M. Lifschitz. *Theoretische Physik, Bd. V*. Akademie Verlag, Berlin, 1979.

[55] C. Lecot. Low discrepancy sequences for solving the Boltzmann equation. *Journal of Computational and Applied Mathematics*, 25:237–249, 1989.

[56] W. Lee and et al. Numerical modeling of advanced semiconductor devices. *IBM J. Res. Develop.*, 36(2):208–230, 1992.

[57] C.D. Levermore. Moment closure hierarchies for kinetic theories. *In preparation*, 1994.

[58] C.D. Levermore and W.J. Morokoff. The Gaussian closure for the Boltzmann equation. *In preparation*, 1994.

[59] T. Makino. Spectroscopic approach for thermal radiation science and in process surface measurement. In W. Nakayama and K. Yang, editors, *Computers and Computing in Heat Transfer Science and Engineering*. CRC Press, 1990.

[60] P. Markowich, C. Ringhofer, and C. Schmeiser. *Semiconductor Equations*. Springer-Verlag, Wien, 1990.

[61] J. P. McVittie, J. C. Rey, A. J. Bariya, M. M. Islamraja, K. Cheng, S. Ravi, and K. C. Saraswat. Speedie: a profile simulator for etching and deposition. In Proc. SPIE Int. Soc. Opt. Eng., editor, *Advanced Techniques for Integrated Processing*, volume 1392, page 126, Santa Clara, 1990, 1991.

[62] J. P. McVittie, J. C. Rey, A. J. Bariya, M. M. Islamraja, S. Ravi, and K. C. Saraswat. Speedie: Simulation of profile evolution during etching and deposition. In *SPIE Symp. Adv. Techniques for Integ. Cts. Processing*, page 126, 1990.

[63] T. P. Merchant, K.-H. Lie, J. V. Cole, and K. F. Jensen. Strategies for modeling of rapid thermal processing systems. In B. Lojek, editor, *Proceedings of the 1st International Rapid Thermal Processing Conference*, page 376, 1993.

[64] B. S. Meyerson. UHV-CVD growth of Si and Si:Ge alloys: Chemistry, physics and device applications. *Proceedings of the IEEE*, 80(10):1592, 1992.

[65] W.J. Morokoff and R.E. Caflisch. A quasi-Monte Carlo approach to particle simulation of the heat equation. *SIAM Num. Anal.*, 30:1558–1573, 1993.

[66] W.J. Morokoff and R.E. Caflisch. Quasi-random sequences and their discrepancies. *SIAM Sci. Comp.*, 15, November, 1994.

[67] W.J. Morokoff and R.E.Caflisch. Quasi-Monte Carlo integration. *Journal of Comp. Phys.*, to appear, 1994.

[68] T. Motohiro. Application of Monte Carlo simulation in the analysis of a sputter deposition process. *J. Vac. Sci. Technol.*, A 4:189, 1986.

[69] Myers. Monte Carlo simulation of sputter atom transport in low pressure sputtering: the effect of interaction potential, sputter distribution and system geometry. *J. Appl. Phys.*, 72(7):3064, 1992.

[70] K. Nanbu. Direct simulation scheme derived from the Boltzmann equation, I: Monocomponent gases. *J. Phys. Soc. Japan*, 49(5):2042–2049, 1980.

[71] K. Nanbu. Variable hard-sphere model for gas mixture. *J. of the Phys. Soc. of Japan*, 59(12):4331, 1990.

[72] K. Nanbu, T. Morimoto, and S. Igarashi. Growth rate of films fabricated with the sputtering method. In *Rarefied Gas Dynamics*, page 913, Aachen, 1990.

[73] H. Niederreiter. Quasi-Monte Carlo methods and pseudo-random numbers. *Bull. Amer. Math. Soc.*, 84:957 – 1041, 1978.

[74] H. Niederreiter. Quasi-Monte Carlo methods for multidimensional numerical integration. In H. Brass and G. Hammerlin, editors, *Numerical Integration III, International Series of Numerical Math. 85*. Birkhauser Verlag, 1988.

[75] H. Niederreiter. *Random Number Generators and Quasi-Monte Carlo Methods, CBMS 63*. Society of Industrial and Applied Mathematics, Philadelphia, 1992.

[76] D.M. O'Brien. Accelerated quasi-Monte Carlo integration of the radiative transfer equation. *J. Quant. Spectrosc. Radiat. Transfer*, 48(1):41–59, 1992.

[77] M. A. Ordal, R. J. Bell, R. W. Alexander, L. L. Long, and M. R. Querry. Optical properties of fourteen metals in the infrared and far infrared. *Applied Optics*, 24(24):4493, 1985.

[78] E. D. Palik, editor. *Handbook of Optical Constants of Solids*. Academic Press, New York, 1985.

[79] S.H. Paskov. Average case complexity of multivariate integration for smooth functions. *Journal of Complexity*, 9(2):291, 1993.

[80] F. Poupaud. On a system of nonlinear Boltzmann equations of semiconductor physics. *SIAM J. Appl. Math.*, 50(6):1593–1606, 1990.

[81] U. Ravaioli. Vectorization of Monte Carlo algorithms for semiconductor simulation. In K. Hess, editor, *Monte Carlo Device Simulation: Full Band and Beyond*, page 267. Kluwer Academic Publishers, Dordrecht, 1991.

[82] J. Rey, L.-Y. Cheng, J. P. McVittie, and K. C. Saraswat. Monte Carlo low pressure deposition profile simulations. *J. Vac. Sci. Technol.*, A 9(3):1083, 1991.

[83] F. Rogier and J. Schneider. A direct method for solving the Boltzmann equation. *Transport Theory and Stat. Phys.*, 23(1-3):313, 1994.

[84] S. M. Rossnagel. Gas density reduction effects in magnetrons. *J. Vac. Sci. Technol.*, A 6(1):19, 1988.

[85] P.K. Sarkar and M.A. Prasad. A comparative study of pseudo and quasi random sequences for the solution of integral equations. *Journal of Comp. Phys.*, 68:66–88, 1987.

[86] S. Selberherr, H. Stippel, and E. Strasser, editors. *Simulation of Semiconductor Devices and Processes, Vol. 1 - 5*. Springer-Velag, Wien, 1993.

[87] R. Siegel and J. R. Howell. *Thermal Radiation Heat Transfer*. Hemisphere Publishing, Washington, 1992.

[88] I.M. Sobol'. The distribution of points in a cube and the approximate evaluation of integrals. *U.S.S.R. Computational Math. and Math. Phys.*, 7:86–112, 1967.

[89] E. Strasser and S. Selberherr. A general simulation method for etching and deposition processes. In E. Strasser S. Selberherr, H. Stippel, editor, *Simulation of Semiconductor Devices and Processes, Vol.5*, page 357. Springer Verlag, 1994.

[90] J. C. Sturm and C. M. Reaves. Silicon temperature measurement by infrared absorption: Fundamental processes and doping effects. *IEEE Trans. on El. Dev.*, 39:81, 1992.

[91] Zhiqiang Tan and P.L. Varghese. The $\delta - \epsilon$ method for the Boltzmann equation. *Journal of Comp. Phys.*, 110(2):327–340, 1994.

[92] B. I. M. ten Bosch, J. J. M. Beenakker, and I. Kuščer. Onsager symmetries in field-dependent flows of rarefied molecular gases. *Physica*, 123 A:443, 1984.

[93] M. W. Thompson. The energy spectrum of ejected atoms during the high energy sputtering of gold. *Philos. Mag.*, 18:377, 1968.

[94] G. M. Turner, I. S. Falconer, B. W. James, and D. R. McKenzie. Monte Carlo calculation of the thermalization of atoms sputtered from the cathode of a sputtering discharge. *J. Appl. Phys.*, 65(9):3671, 1989.

[95] Th. Vogelsang. *Monte Carlo Simulation des Transports heisser Elektronen im Silizium*. PhD thesis, Technische Universität München, München, 1994.

[96] W. Wagner. A convergence proof for Bird's direct simulation Monte Carlo method for the Boltzmann equation. *Communications on Mathematical Physics*, 1991.

[97] L. Waldmann. Transporterscheinungen in Gasen von mittlerem Druck. In *Handbuch der Physik Vol.XII*. Springer Verlag, 1958.

[98] G. J. Ward. Measuring and modeling anisotropic reflection. *Computer Graphics*, 26(2), 1992.

[99] W. D. Westwood. Physical vapor deposition. In R. A. Levy, editor, *Microelectronic Materials and Processing*. Kluwer Academic Publishers, 1989.

[100] H. Wozniakowski. Average case complexity of multivariate integration. *Bull. Amer. Math. Soc.*, 24:185–194, 1991.

[101] H. C. Wulu, K. C. Saraswat, and J. P. McVittie. Simulation of mass transport for deposition in via holes and trenches. *J. Electrochem. Soc.*, 138(6):1831, 1991.

Index

Progress in Numerical Simulation for Microelectronics

R. Kircher / W. Bergner
SIEMENS AG, München,
Germany

Three-Dimensional Simulation of Semiconductor Devices

1991. 128 pages. Hardcover
ISBN 3-7643-2644-1

Please order through your bookseller or write to:
Birkhäuser Verlag AG
P.O. Box 133
CH-4010 Basel / Switzerland
FAX: ++41 / 61 / 271 76 66

For orders originating in the USA or Canada:
Birkhäuser
333 Meadowlands Parkway
Secaucus, NJ 07094-2491 / USA

Birkhäuser

Birkhäuser Verlag AG
Basel · Boston · Berlin

Prices are subject to change without notice 5/94

In the last two decades the simulation of semiconducting device structures has become an increasingly important tool for developing and improving devices by predicting their electrical behaviour. Today the two-dimensional device simulation is regarded as a standard engineering tool, containing sophisticated models describing transport phenomena. The challenge arising from the increasing integration density is to consider the influence of a more and more three-dimensional geometry of the devices.

Insight is given here into modeling complex device structures in three dimensions. Starting from the basic differential equations and the physical models describing the transport in semiconductors, the work shows how to obtain the discretized equations, and discusses important numerical methods for solving these equations. How three-dimensional simulations can be utilized to analyze complex device structures and to optimize their electrical characteristics is illustrated by two specific examples.

Progress in Numerical Simulation for Microelectronics

C.R. Kleijn, TU Delft, The Netherlands / **C. Werner,** Siemens ZFE, München, Germany

Modeling of Chemical Vapor Deposition of Tungsten Films

1993. 138 pages. Hardcover
ISBN 3-7643-2858-4

Please order through your bookseller or write to:
Birkhäuser Verlag AG
P.O. Box 133
CH-4010 Basel / Switzerland
FAX: ++41 / 61 / 271 76 66

For orders originating in the USA or Canada:
Birkhäuser
333 Meadowlands Parkway
Secaucus, NJ 07094-2491 / USA

Birkhäuser

Birkhäuser Verlag AG
Basel · Boston · Berlin

Prices are subject to change without notice. 5/94

Numerical modeling of reactors for chemical vapor deposition has in recent years become a field of great interest, because it offers the potential to support development and optimization of manufacturing equipment and hence reduce the cost and improve the quality of the reactors.

This book is the result of two parallel lines of research dealing with the same subject – Modeling of Tungsten CVD Processes –, which were performed independently under very different boundary conditions. On the one side, Chris Kleijn, working in the adacemic research environment of Technical University Delft, was able to go deep enough into the subject to lay a solid foundation and prove the validity of all the assumptions made in his work. On the other side, Christoph Werner, working in the context of an industrial research lab at Siemens corporate research and development, was able to closely interact with manufacturing and development engineers in a modern submicron semiconductor processing line.

ISNM 106 **S.N. Antontsev, K.-H. Hoffmann, A.M. Khludnev (Eds):** Free Boundary
Problems in Continuum Mechanics. 1992 (ISBN 3-7643-2784-7)

ISNM 107 **V. Barbu, F.J. Bonnans, D. Tiba (Eds):** Optimization, Optimal Control and
Partial Differential Equations. 1992 (ISBN 3-7643-2788-X)

ISNM 108 **H. Antes, P.D. Panagiotopoulos:** The Boundary Integral Approach to Static and
Dynamic Contact Problems. Equality and Inequality Methods.
1992 (ISBN 3-7643-2592-5)

ISNM 109 **A.G. Kuz'min:** Non-Classical Equations of Mixed Type and their Applications in
Gas Dynamics. 1992 (ISBN 3-7643-2573-9)

ISNM 110 **H.R.E.M. Hörnlein, K. Schittkowski (Eds):** Software Systems for Structural
Optimization. 1992 (ISBN 3-7643-2836-3)

ISNM 111 **R. Burlisch, A. Miele, J. Stoer, K.H. Well:** Optimal Control.
1993 (ISBN 3-7643-2887-8)

ISNM 112 **H. Braess, G. Hämmerlin (Eds):** Numerical Integration IV. Proceedings
of the Conference at the Mathematical Research Institute at Oberwolfach,
November 8–14, 1992. 1993 (ISBN 3-7643-2922-X)

ISNM 113 **L. Quartapelle:** Numerical Solution of the Incompressible Navier-Stokes
Equations. 1993 (ISBN 3-7643-2935-1)

ISNM 114 **J. Douglas, U. Hornung (Eds):** Flow in Porous Media.
1993 (ISBN 3-7643-2949-1)

ISNM 115 **R. Bulirsch, D. Kraft (Eds):** Computational Optimal Control.
1994 (ISBN 3-7643-5015-6)

ISNM 116 **P.W. Hemker, P. Wesseling (Eds):** Multigrid Methods IV. Proceedings of the
Fourth European Multigrid Conference, Amsterdam, July 6-9, 1993.
1994 (ISBN 3-7643-5030-X)

ISNM 117 **R.E. Bank, R. Bulirsch, H. Gajewski, K. Merten:** Mathematical Modelling and
Simulation of Electrical Circuits and Semiconductor Devices. 1994
(ISBN 3-7643-5053-9)

ISNM 118 **W. Desch, F. Kappel, K. Kunisch (Eds):** Control and Estimation of Distributed
Parameter Systems: Nonlinear Phenomena. International Conference on Control
and Estimation of Distributed Parameter Systems, Vorau, July 18-24, 1993.
1994 (ISBN 3-7643-5098-9)